Laboratory Manual

Activities·Experiments·Demonstrations·Tech Labs

Paul G. Hewitt and Dean Baird

CONCEPTUAL Physics
eleventh edition

Paul G. Hewitt

City College of San Francisco

Addison-Wesley

Boston Columbus Indianapolis New York San Francisco Upper Saddle River
Amsterdam Cape Town Dubai London Madrid Milan Munich Paris Montréal Toronto
Delhi Mexico City São Paulo Sydney Hong Kong Seoul Singapore Taipei Tokyo

Publisher: Jim Smith
Director of Development: Michael Gillespie
Editorial Manager: Laura Kenney
Project Editor: Chandrika Madhavan
Director of Marketing: Christy Lawrence
Senior Marketing Manager: Kerry Chapman
Managing Editor: Corinne Benson
Production Supervisor: Mary O'Connell
Production Service: Progressive Publishing Alternatives
Cover Photo: NASA
Supplement Cover Designer: Seventeenth Street Studios
Text and Cover Printer: Edwards Brothers
Manufacturing Buyer: Jeff Sargent

ISBN 10: 0-321-73248-0
ISBN 13: 978-0-321-73248-4

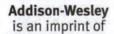

Addison-Wesley
is an imprint of

www.pearsonhighered.com

1 2 3 4 5 6 7 8 9 10 – EB – 15 14 13 12 10

Introduction

Many students enter this course having had very few hands-on science learning experiences. The activities, experiments, tech labs, and demonstrations in this manual are designed to provide the experiences students need. Whether it's using a pinhole image to determine the size of the Sun, using probeware to plot real-time motion graphs, or attempting to use a battery and wires to light a bulb, these explorations are designed to put students in direct interaction with the natural world in ways that will deepen both their curiosity and their understanding.

This manual likely has more activities than you will have time to complete. The instructor of the course is in the best position to decide which activities to do and which to leave out. Some of these decisions may be made for you based on apparatus availability or constraints. Students will be well served if they are given a broad exposure to activities across the range of content coverage, activity type, and difficulty level.

It has been suggested that we retain about 5% of what we read and about 75% of what we do. So the activities, experiments, demonstrations, and tech labs in this manual play an important role in the complete *Conceptual Physics* program of instruction. Enjoy!

Acknowledgments

A great number of influences came together in the activities written for this lab manual. Dean's very first professional influence was that of his mentor, Walter Scheider, at Ann Arbor Huron High in Michigan.

We are grateful to our colleagues in the American Association of Physics Teachers (AAPT) for a never-ending stream of wonderful ideas, techniques, and demonstrations that spur our creativity. Physics instructors are a clever bunch, and that cleverness is on display at national meetings of the AAPT. Dewey Dykstra and Jim Minstrell were Dean's important early AAPT influences.

The Northern California and Nevada section of the AAPT is home to a number of exceptional physics instructors, whose contributions at meetings we have benefited. Clarence Bakken, Dan Burns, Ann Hanks, and David Kagan have provided invaluable guidance over the years.

San Francisco's famous Exploratorium hands-on science museum is a shining beacon of science and demonstrations of perception not only in northern California, but worldwide. Paul Doherty and Dan Rathjen shower area physics teachers with physics teaching gems that are simple and effective. Their energy is in this manual, too.

Dean's PTSOS colleagues Dan Burns, Stephanie Finander, and Steve Keith have provided surprising and novel ways to present physics topics in engaging ways. For lab ideas, we are grateful to PTSOS leader Paul Robinson of San Mateo High School, San Mateo, California; as well as Earl R. Feltyberger of Nicolet High School in Glendale, Wisconsin; Jessica Downing of Natomas, California; and retired physics teacher Howie Brand. For final manuscript tweaking, we thank Lillian Lee Hewitt.

At Rio Americano, we are particularly grateful to physics teacher Lucy Jeffries for her patience with Dean's continual experimentation with experiments and for her feedback on labs, as well as her willingness to venture together into the choppy waters of computer-based lab activities!

Physics students at Rio Americano are the inspiration for much of this manual. They have also been the most important source of feedback for revisions of all these activities, long before they were included in this manual. Small groups of Rio's physics students even played a role in editing early versions of the manuscript. It is a credit to their English instructors and to their own mastery of the language that they could spot the tiniest of mistakes— even in the small print!

From Dean's point of view, no one has had a greater influence on his physics teaching than coauthor, Paul Hewitt. As Dean says, "Paul's work throughout the years has been the most important star in the constellation by which I steer the ship of my curriculum."

From Paul's point of view, no one is better able to connect with students in the laboratory than Dean Baird, which is why Paul chose Dean to coauthor this manual—which is delightfully more Baird than Hewitt!

Table of Contents

About Paul G. Hewitt

Paul G. Hewitt, former boxer, uranium prospector, sign painter, and cartoonist began college at the age of 28 and fell in love with physics. Hewitt's teaching career began in 1964 at City College of San Francisco. In 1971, he authored the first edition of *Conceptual Physics*, a physics textbook for nonscientists that revolutionized physics instruction.

Hewitt has taken leaves to teach physics at the University of California, both at the Berkeley and Santa Cruz campuses, and at the University of Hawaii at both the Hilo and Manoa campuses. He taught an evening course for the general public at the Exploratorium in San Francisco. The Exploratorium honored Hewitt with its Outstanding Educator Award in 2000.

In recognition of Hewitt's achievements, the American Association of Physics Teachers (AAPT) honored him with their Millikan Award for outstanding contributions to physics teaching.

Hewitt produces and illustrates the cartoon column, "Figuring Physics," for *The Physics Teacher,* the monthly magazine of the AAPT.

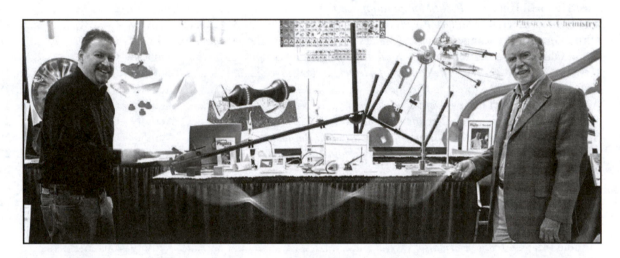

Dean Baird and Paul Hewitt

About Dean Baird

Dean Baird, former paperboy and grocery bagger, began college at the age of 17 and fell in love with physics. Baird's teaching career began in 1986 at Rio Americano High School in Sacramento. In 1986, he began authoring *The Book of Phyz* and in 2000, he posted this comprehensive curriculum resource to his website at http://phyz.org.

Baird has acted as mentor in the NSF-funded Southern California Area Mentors of Physics Instruction, served on the AAPT's Examinations Editorial Board and as an officer in the Northern California/Nevada American Association of Physics Teachers (NCNAAPT). He was also a member of the Golden State Examination Development Team.

In recognition of Baird's accomplishments, the NCNAAPT honored him with a Distinguished Service Citation. The AAPT also honored him with its Distinguished Service Citation.

Baird serves as an appointed member of the California Department of Education's Assessment Review Panel. He is also a workshop leader for PTSOS, a program that supports new physics teachers. He is the author of lab manuals for *Conceptual Physical Science* and *Conceptual Integrated Science*.

To the Student

Just as you need to know the rules of a game to fully appreciate it—whether sports, a party game, or a computer game—so it is with nature. You'll have a fuller appreciation of nature when you know something about how things connect to one another and how physical processes follow certain rules. The natural world about you is more comprehensible when you know something about its rules.

This *Conceptual Physics Laboratory Manual* adds to that by focusing on how nature keeps score. Your laboratory experience is a hands-on experience, where *you* make the observations, *you* make the measurements, *you* make the calculations, and *you* draw the conclusions. Taken together, lab work makes real the principles and relationships you learn in the classroom and read about in your textbook. By the way, be sure to bring your textbook with you to lab! There are sometimes references to textbook diagrams that might help you understand what's going on during a lab activity.

To the Instructor

Doing physics is what this lab manual is about. Inside is a series of activities, experiments, tech labs, and demonstrations that are closely tied to the eleventh edition of *Conceptual Physics*.

The great number and variety of laboratory activities included in this manual is intended to allow you to pick and choose as you prepare your lab sessions. You'll likely be guided by available apparatus and time constraints. But the point is to give you an adequate selection from which to choose.

Activities are often qualitative and relatively quick to do. They may involve prediction, observation, and deduction. Responses are usually verbal, but may involve simple sketches, too. In some cases, activities simply set the stage for particular chapter material; in other cases, they summarize concepts in the textbook.

Experiments mostly involve making measurements, and making both qualitative and quantitative assessments. Experiments may extend the chapter material, and some require a bit of algebra and some a bit of trigonometry.

Tech Labs involve the use of computers. Some are probeware-based, requiring sensors that interface with the computer as well as data-collection and analysis software. Pasco's PasPort sensors and *DataStudio* software, and Vernier's detectors and *Logger Pro* software are leaders in the field. While handheld data collection tools are becoming increasingly popular, computer-interfaced probeware allows lab groups to gather around one relatively large screen and discuss results as a group. Some tech labs are computer simulations. The University of Colorado's Physics Education Technology Group (PhET) is continually developing robust computer simulations. While some simulation-based tech labs *might* work as outside assignments, they are much more valuable when worked on by a group. Group discussions during these activities have great benefits for all involved.

Demonstrations are activities that simply aren't viable as lab group activities. For example, few institutions have one Van de Graaff generator for each lab group. Unlike lecture demonstrations, which briefly illustrate a fairly concise point, lab demonstrations are intended as extended lessons based on the demonstration itself. Student involvement is encouraged for most demonstrations, but demonstrations are usually more instructor-led than student-led. You may be able to do some demonstrations as activities or stations, depending on your institution's resources.

Notice that the lab manual is tied to the textbook. The table of contents listing for each lab gives its textbook chapter reference. That textbook reference is repeated near the top-left of the actual lab write-up. Figures and other specific references to the textbook arise from time to time in the laboratory; encourage students to bring they're textbook to lab!

Any observations you might have about the activities in this manual are welcome by the authors. Please email your comments, critiques, and suggestions directly to Dean Baird, dean@phyz.org.

Master List of Apparatus

Item	Activities and Experiments
acetate strips	*A Force to be Reckoned, Electroscopia*
acrylic envelope	*Seeing Magnetic Fields*
adhesive tape	*It's All in the Wrist, Laser Tree*
air core solenoid	*Bobbing for Magnets, Generator Activator*
Air-Powered Projectile (Arbor)	*Blast Off!*
air pump	*Blast Off!*
aluminum cake pan, 9"	*Boat Float*
aluminum foil	*I'm Melting, I'm Melting, The Lemon Electric (Optional), Pinhole Camera*
ammeter, DC	*Ohm Ohm on the Range, An Open and Short Case*
aquarium tank	*Sink or Swim*
automobile	*Tire Pressure and 18-Wheelers*
balance (electronic)	*Making Cents, Blowout! (Optional), The Fountain of Fizz, Blast Off!, Eureka!, Sink or Swim, Boat Float*
balance, equal arm	*Spiked Water*
ball, steel	*The Newtonian Shot, Dropping the Ball, Bull's Eye*
ballast (e.g., sand, BBs, nuts)	*Eureka!*
balloons (2)	*Charging Ahead*
barium chloride powder	*Bright Lights*
batteries, C-cell	*The Lemon Electric, Batteries and Bulbs*
batteries, D-cell	*The Lemon Electric, Batteries and Bulbs, An Open and Short Case, Motor Madness*
battery, dry ignition cell (No. 6)	*The Lemon Electric*
battery, lantern (6-volt)	*Electric Magnetism, Motor Madness (Optional), Bobbing for Magnets*
battery, N- or AAAA-cell	*The Lemon Electric*
BB shot	*The Big BB Race, Thickness of a BB Pancake*
beaker	*Slow-Motion Wobbler*
beaker, 400-mL	*Cooling by Boiling*
bicycle wheel (with handles)	*Sit on It and Rotate*
boat, toy	*Boat Float*
book (heavy)	*A Force to Be Reckoned*
bottled water (1/2 liter plastic)	*It's All in the Wrist (Optional)*
bowl	*Sink or Swim*
bricks	*A Force to Be Reckoned*
bubble solution and wand or bubble gun	*Charging Ahead*
bucket, 3-gallon	*Eureka!, Boat Float, Temperature Mix*
bulb sockets, miniature	*Batteries and Bulbs, An Open and Short Case, Be the Battery*
bulbs, miniature	*Ohm Ohm on the Range, Ohm Ohm on the Digital Range, Resistance is Not Futile, Batteries and Bulbs, An Open and Short Case, Be the Battery*
Bunsen burner	*Bright Lights*
calcium chloride powder	*Bright Lights*
can, empty	*Bull's Eye*
candle	*Tuning the Senses*
cans, radiation (silver, black, white)	*Canned Heat: Heating Up, Canned Heat: Cooling Down*
cardboard	*Sunballs*
cards, 3" x 5" (index)	*Dance of the Molecules (Optional), Pinhole Image, Image of the Sun*
catch box	*Blowout!*
CD	*Electric Magnetism*
chair seat, plastic	*A Force to Be Reckoned (Optional), Electroscopia*
chalkboard and chalk	*Making Cents*
chalk dust	*Oleic Acid Pancake*
charged sphere	*Charging Ahead (Optional)*
clamp, right angle	*The Big BB Race, Bobbing for Magnets, Pole-Arizer*
clamps, three-finger or buret	*Dropping the Ball, Totally Stressed Out*
clay, modeling	*Boat Float*

collar hooks	*Walking the Plank, The Weight, Totally Stressed Out, Motor Madness, Bobbing for Magnets*
compact disk/cassette tape	*Sound Off*
compact fluorescent light	*Diffraction in Action*
compasses, navigational (4)	*Electric Magnetism*
computer	*Making Cents, Sonic Ranger, Putting the Force Before the Cart, The Force Mirror, Blast Off!, Worlds of Wonder, Spring to Another World, Water Waves in an Electric Sink, High Quiet Low Loud, Wah-Wahs and Touch Tones, Electric Field Hockey, Bicycle Dancer of Edinburgh, Bouncing Off the Walls*
container	*Get a Half-Life*
crossbars (short rods)	*Walking the Plank, The Weight, Totally Stressed Out, Bobbing for Magnets, Pole-Arizer*
cupric chloride powder	*Bright Lights*
dart guns (2) with darts (2)	*The Newtonian Shot*
defrosting trays	*I'm Melting, I'm Melting*
diffracton ("rainbow") glasses	*Laser Tree (optional)*
diffraction grating, any	*Diffraction in Action, Bright Lights*
diffraction grating, ~ 500 lines/mm & ~ 1000 lines/mm	*Laser Tree*
digital multimeter	*The Lemon Electric*
dime	*Sunballs*
dollar bill	*Reaction Time*
dominoes (100)	*Chain Reaction*
dowel, wood	*Motor Madness*
dumbbells (2)	*Sit on It and Rotate*
dynamics cart and track	*Putting the Force Before the Cart, An Uphill Climb, Will it Go 'Round in Circles?*
eggs, raw	*Egg Toss, Sink or Swim*
electric drill	*Bouncy Board*
electrophorus	*A Force to Be Reckoned (Optional), Electroscopia*
electroscope	*Electroscopia*
ethanol (995 mL)	*Oleic Acid Pancake*
eyedropper	*Oleic Acid Pancake*
film canisters (35 mm) or small, plastic bottles with string attached	*Eureka!*
floor space or flat surface	*Get a Half-Life, Chain Reaction*
fog-in-a-can	*Laser Tree (Optional)*
food coloring (2 colors)	*Dance of the Molecules*
funnel	*The Fountain of Fizz*
galvanometer	*Generator Activator*
galvanometer, demonstration	*Bobbing for Magnets*
garbage bags (2)	*Egg Toss*
gas discharge tubes	*Bright Lights*
gauge, tire pressure	*Tire Pressure*
generator, hand-crank	*Bobbing for Magnets*
generator, handheld	*Be the Battery, Motor Madness, Generator Activator*
graduated cylinder, 500-mL	*Eureka!, Boat Float*
graduated cylinder, 250-mL & matching length of PVC tube	*Fork it Over*
graduated cylinder, 100-mL	*Thickness of a BB Pancake*
graduated cylinder, 10-mL	*Oleic Acid Pancake*
graduated cylinder, wide-mouth	*Sink or Swim*
HCl (hydrochloric acid)	*Bright Lights*
hair dryer	*Electroscopia (Optional)*
headphones	*Bicycle Dancer of Edinburgh, High Quiet Low Loud, Wah-Wahs and Touch Tones*
heat-generating pouch (sodium acetate pack)	*Heating by Freezing*
heat lamp and base	*Canned Heat: Heating Up*
heat-proof glove or potholder	*Dance of the Molecules (Optional), Bright Lights*
hex nuts (4)	*Putting the Force Before the Cart*
hot plate	*Cooling by Boiling, Heating by Freezing*

ice cubes (similar in size)	*I'm Melting, I'm Melting*
interface device with connectors	*Sonic Ranger, Ohm Ohm on the Digital Range*
interface device for sensors	*Putting the Force Before the Cart, The Force Mirror, Ohm Ohm on the Digital Range, Resistance is Not Futile*
Internet access	*Bicycle Dancer of Edinburgh, Wah-Wahs and Touch-Tones (Optional), Resistance is Not Futile*
iron filings	*Seeing Magnetic Fields*
jars, baby food (2, empty)	*Dance of the Molecules*
knife (small, or X-acto blade)	*Image of the Sun, Sunballs*
lab partner	*Reaction Time, The Force Mirror, Sit on it and Rotate, The Big BB Race, Be the Battery, Mirror rorriM*
laser or laser pointer, any	*A Sweet Mirage, Trapping the Light Fantastic, Light Rules (with known wavelength*
laser pointer, green, blue, or violet	*Laser Tree*
laser pointer, red	*Laser Tree*
Laser Viewing Tank (Arbor)	*Will it Go 'Round in Circles? (Optional), Why the Sky is Blue, Trapping the Light Fantastic, A Sweet Mirage, Laser Tree*
Launchpad (Arbor)	*Blast Off!*
lead blocks	*Sink or Swim*
lead weights (or fishing weights)	*Boat Float*
LED Mini Maglite	*Why the Sky is Blue, Diffraction in Action (incandescent light & white light [Optional])*
lemon half or wedge	*The Lemon Electric*
lens, 25 mm converging	*Pinhole Camera*
light source, bright (e. g., analog projector or OHP)	*Blackout*
liter container	*Temperature Mix (1-L wide mouth [Optional]), Trapping the Light Fantastic (2-L clear plastic [Optional]),*
lycopodium powder	*Oleic Acid Pancake*
magnet, small neodymium	*Dropping the Ball, Bobbing for Magnets (Optional)*
magnetic field projectual	*Seeing Magnetic Fields*
magnets, bar	*A Force to be Reckoned, Motor Madness, Bobbing for Magnets, Seeing Magnetic Fields, Generator Activator*
mallet, tuning fork	*Slow-Motion Wobbler, Fork it Over, Why the Sky is Blue*
mass blocks	*Putting the Force Before the Cart, Boat Float (100-g)*
mass hanger	*The Weight, It's All in the Wrist, Totally Stressed Out*
masses, hooked	*Bouncy Board, It's All in the Wrist, Boat Float*
masses, slotted	*Walking the Plank, The Weight, It's All in the Wrist, Totally Stressed Out*
masses (various)	*Bouncy Board*
matches	*Tuning the Senses, Charging Ahead (wooden)*
Mentos (7 candies)	*The Fountain of Fizz*
metal, short rod or tube	*Electroscopia*
meterstick	*Walking the Plank, Go! Go! Go!, Bouncy Board, An Uphill Climb, The Fountain of Fizz, Dropping the Ball, Twin-Baton Paradox, It's All in the Wrist, Bull's Eye, Blast Off!, Totally Stressed Out, Fork it Over, Pinhole Image, Image of the Sun, Sunballs, Light Rules*
meterstick clamp	*It's All in the Wrist, Totally Stressed Out*
micrometer	*Thickness of a BB Pancake*
mirror, full length	*Mirror rorriM*
mirror, pocket (Optional)	*Mirror rorriM*
nails (galvanized) (2)	*The Lemon Electric*
nails (short, tied with string)	*Spiked Water*
Newton's cradle	*Will it Go 'Round in Circles? (Optional)*
nichrome flame test wires	*Bright Lights*
oleic acid (5 mL)	*Oleic Acid Pancake*
opaque white tank insert	*Trapping the Light Fantastic*
paper/notebook	*Tuning the Senses, Spring to Another World, Fork it Over (Optional), Seeing Magnetic Fields, Image of the Sun, Blackout (Optional)*

paper towels	*Spiked Water, Canned Heat: Heating Up, Canned Heat: Cooling Down, I'm Melting, I'm Melting, Fork it Over, The Lemon Electric, Why the Sky is Blue*
paper, butcher	*Go! Go! Go!*
paper, graph	*Making Cents, Go! Go! Go!, The Weight, Totally Stressed Out, Tire Pressure and 18-Wheelers, Canned Heat: Heating Up, Canned Heat: Cooling Down, Ohm Ohm on the Range*
paper clip	*Putting the Force Before the Cart*
paper, tracing	*Pinhole Camera*
pen/pencil	*Tuning the Senses, Spring to Another World, Image of the Sun, Sunballs, Diffraction in Action*
pencils, colored	*Bright Lights, Get a Half-Life*
pencil with eraser end	*Go! Go! Go!*
pennies	*Making Cents, The Lemon Electric (Optional)*
pens, black felt-tip	*Go! Go! Go!, Temperature Mix, I'm Melting, I'm Melting*
pen, marker	*Blowout!*
photogate & timer	*Blowout! (Optional), Dropping the Ball*
pie tins, small	*A Force to Be Reckoned (Optional), Charging Ahead*
pins, straight	*Pinhole Image*
pipe, aluminum (5 ft)	*It's All in the Wrist (Optional)*
pith balls & string	*A Force to be Reckoned*
plastic objects (transparent)	*Blackout*
plastic plate	*The Lemon Electric*
platform	*Electric Magnetism*
playing cubes, 25 (dice or painted sugar cubes)	*Get a Half-life*
playing field	*Egg Toss, Blast Off!*
polarizer, large (with diffuser) & small (filter)	*Blackout*
polarizing film (large sheet)	*Blackout*
polypropylene ropes, 2	*Egg Toss*
pot, large	*Heating by Freezing*
potassium chloride powder	*Bright Lights*
power supply, variable DC (0–6 V)	*Ohm Ohm on the Range, Ohm Ohm on the Digital Range, Resistance is Not Futile*
pressure gauge, tire	*Tire Pressure and 18-Wheelers*
printer access	*Ohm Ohm on the Digital Range, Resistance is Not Futile*
protractor	*An Uphill Climb*
pulley	*Putting the Force Before the Cart*
PVC tube	*Fork it Over*
resistance spools or coils	*Resistance is Not Futile*
resistors, power	*Ohm Ohm on the Range, Ohm Ohm on the Digital Range*
ring clamp	*Electric Magnetism*
ring stand with large clamp	*Bright Lights*
rock (or hooked mass)	*Boat Float*
rod clamps	*Walking the Plank, The Weight, An Uphill Climb, Motor Madness*
rotational batons (1 pair)	*Twin-Baton Paradox*
rotational stool	*Will it Go 'Round in Circles? (platform), Sit on It and Rotate*
rubber bands	*The Force Mirror (#64), Totally Stressed Out (#64 [Optional]), Motor Madness*
ruler, centimeter	*Reaction Time, Thickness of a BB Pancake, Mirror, rorriM,*
ruler, etched metal (mm)	*Light Rules*
safety glasses	*Egg Toss*
salt	*Sink or Swim*
scattering agent	*Why the Sky is Blue, Trapping the Light Fantastic, A Sweet Mirage, Laser Tree*
sensor, current	*Ohm Ohm on the Digital Range, Resistance is Not Futile*
sensor, force	*The Force Mirror*
sensor, motion	*Sonic Ranger, Putting the Force Before the Cart*
sensor, voltage	*Ohm Ohm on the Digital Range, Resistance is Not Futile*
shoebox with lid	*Pinhole Camera*
signal splitters	*Bicycle Dancer of Edinburgh, High Quiet Low Loud, Wah-Wahs and Touch Tones*

silk cloth square	*A Force to be Reckoned, Electroscopia*
simultaneous launcher	*The Big BB Race*
soda pop (2 aluminum cans of regular and diet)	*Sink or Swim*
soda pop (diet, 2-L bottle)	*The Fountain of Fizz*
sodium chloride powder	*Bright Lights*
software, analysis	*Making Cents, Ohm Ohm on the Digital Range, Resistance is Not Futile*
software for graphing motion & force sensor data	*Sonic Ranger, Putting the Force Before the Cart, The Force Mirror*
software, PhET simulations	*Blast Off!, Worlds of Wonder, Spring to Another World, Bouncing Off the Walls, Water Waves in an Electric Sink, Electric Field Hockey*
software, sound-generating (such as Pasco's DataStudio with WavePort)	*High Quiet Low Loud, Wah-Wahs and Touch Tones*
solder, lead-free (~50 cm)	*Motor Madness*
spatula (metal)	*Bright Lights*
spectroscope	*Bright Lights*
spoon	*Sink or Swim, Temperature Mix*
spirit level (small)	*Walking the Plank (Optional)*
spring, weak	*Totally Stressed Out, Bobbing for Magnets*
spring scales	*Walking the Plank, The Weight, The Force Mirror, An Uphill Climb, Twin-Baton Paradox, Boat Float*
stereo audio device with 2 moveable speakers	*Sound Off*
stirring rod	*Why the Sky is Blue, Trapping the Light Fantastic, A Sweet Mirage, Laser Tree*
St. Louis Motor	*Motor Madness (Optional)*
stopwatch	*Go! Go! Go!, The Fountain of Fizz, Bull's Eye, Dance of the Molecules, Canned Heat: Heating Up, Canned Heat: Cooling Down, Chain Reaction*
string	*Walking the Plank, Putting the Force Before the Cart, It's All in the Wrist (Optional), Eureka!, Bouncy Board, Spiked Water, A Force to Be Reckoned*
strobe light, variable frequency	*Slow-Motion Wobbler*
strontium chloride powder	*Bright Lights*
Styrofoam blocks	*Sink or Swim, Electroscopia*
Styrofoam bowls	*Charging Ahead*
Styrofoam cups	*Temperature Mix, Spiked Water*
Styrofoam plates (white & black)	*I'm Melting, I'm Melting*
sugar cubes	*A Sweet Mirage, Get a Half-Life*
sunlight (direct)	*Image of the Sun*
support rods	*Walking the Plank, The Weight, The Force Mirror, Blowout!, An Uphill Climb, Dropping the Ball, The Big BB Race, Totally Stressed Out, A Force to Be Reckoned, Electric Magnetism (with base), Motor Madness (with base & clamp), Bobbing for Magnets (with base), Pole-Arizer (4 m)*
support base (or table clamp)	*Totally Stressed Out, A Force to Be Reckoned, Electric Magnetism, Motor Madness, Bobbing for Magnets*
switch, double pole double throw	*Sound Off*
table	*Go! Go! Go!, The Force Mirror, Bouncy Board, An Uphill Climb, Dropping the Ball, Bull's Eye, I'm Melting, I'm Melting, Slow-Motion Wobbler, Seeing Magnetic Fields, Chain Reaction*
table clamps	*Walking the Plank, The Weight, The Force Mirror, An Uphill Climb, Dropping the Ball, The Big BB Race, Totally Stressed Out*
tape, masking	*Go! Go! Go!, Blowout!, Egg Toss, Eureka!, Boat Float, Pinhole Camera, Mirror, rorriM,*
tape measure	*Egg Toss (100-ft), Blast Off! (300-ft), Light Rules*
telephone cord, 25-ft coiled	*Pole-Arizer*
thermometer (Celsius)	*Temperature Mix, Spiked Water, Canned Heat: Heating Up, Canned Heat: Cooling Down, Cooling by Boiling, Fork it Over*
thrust washers	*Blast Off!*
timer	*Tuning the Senses*
toy car, constant velocity	*Go! Go! Go!*
tray	*Thickness of a BB Pancake, Oleic Acid Pancake*

trundle wheel	*Egg Toss, Blast Off!*
tube, copper or aluminum	*Bobbing for Magnets (Optional)*
tube, 5- to 10-ft	*Blowout!*
tube, acrylic, 4-ft	*Dropping the Ball*
tuning forks	*Blast Off! (300 Hz–500 Hz), Slow Motion Wobbler (low frequency), Fork it Over (unknown & 300 Hz–500 Hz)*
tuning forks (2) attached to wooden sound boxes	*Why the Sky is Blue*
vacuum pump with bell jar	*Cooling by Boiling*
Van de Graaff generator	*Charging Ahead*
vehicle owner's manual	*Tire Pressure and 18-Wheelers*
vinyl strips	*A Force to be Reckoned, Electroscopia*
Visual Accelerometer (Pasco) (2)	*Will it Go 'Round in Circles?*
voltmeter, DC (analog)	*Ohm Ohm on the Range*
volunteers	*Will it Go 'Round in Circles?, Pinhole Image (1 nearsighted & 1 farsighted), Pole-Arizer,*
water	*Oleic Acid Pancake, Eureka!, Sink or Swim, Boat Float, Dance of the Molecules (hot & cold), Temperature Mix (hot & cold), Spiked Water (hot & cold), Canned Heat: Heating Up (cold), Canned Heat: Cooling Down (hot), Heating by Freezing (boiling), Slow-Motion Wobbler, Fork it Over, The Lemon Electric, Why the Sky is Blue, Trapping the Light Fantastic, A Sweet Mirage (hot), Laser Tree*
wire, coiled into many loops	*Generator Activator*
wires, connecting	*Ohm Ohm on the Range, Ohm Ohm on the Digital Range, Resistance is Not Futile, Batteries and Bulbs, An Open and Short Case, Be the Battery, Electric Magnetism, Motor Madness, Bobbing for Magnets, Generator Activator*
wood angle wedges (Arbor)	*Blast Off!*
wood blocks	*Putting the Force Before the Cart, Sink or Swim, Boat Float, Motor Madness*
wool cloth squares	*A Force to be Reckoned, Electroscopia*
YouTube video v=Z19zFlPah-o	*Bicycle Dancer of Edinburgh*

A Partial List of Vendors and Their URLs

Arbor Scientific · www.arborsci.com

Sargent-Welch · www.sargentwelch.com

Vernier · www.vernier.com

Pasco Scientific · www.pasco.com

Science Kit · www.sciencekit.com

Name _____ Section _____ Date _____

CONCEPTUAL PHYSICS	Activity

Tuning the Senses

Purpose
To tune your senses of sight and sound

Scientists' original source of information about the universe comes from personal observations. This leads to questioning reasons and causes. A scientist notices something, asks questions, and then tries to answer them. By this definition, can't we all be scientists?

Apparatus
notebook, pen or pencil, candle, matches, clock or timer, and patience

Discussion
Galileo wrote, "In questions of science the authority of a thousand is not worth more than the humble observation and reasoning of a single individual." You'll do two simple activities, the first to tune your hearing, the second to tune your seeing.

Procedure
Perform the following activities on your own and then answer the questions on the following page.

PART A: AUDITION OF THE ENVIRONMENT
Go outside and find a comfortable place to sit. Listen to your environment for 10 minutes. Write down the sounds you hear. You might find it helpful to close your eyes during this activity—don't let the sights of your surroundings distract you from observing its sounds.

PART B: OBSERVATION OF A BURNING CANDLE
Remain absolutely quiet while observing an unlit candle.

Light the candle, observe it for 2 minutes, then record your observations.

After you have exhausted all observations you can think of, extinguish the candle.

Observe the extinguished candle while recording more observations.

What other activities can you invent to tune your senses?

Be creative and tune your senses every day!

Summing Up Part A: Audition of the Environment

1. What was the quietest sound you heard? What was the loudest?

_____

2. Which sounds had a relatively high pitch (a chirping bird, for example)?

_____

3. Which sounds had a relatively low pitch (a truck engine, for example)?

_____

4. How many sounds were natural, and how many were made by humans (or human activity)?

_____

5. Did you identify sounds by their sources, or did you spell them out phonetically?

_____

6. Are there any continuous sounds *right now* (like that of an air conditioner) that you have simply tuned out? What are they?

_____

7. How might you describe a sound to someone who was unfamiliar with its source (for example, the sound of a car to someone who had never heard of a car before)?

_____

Summing Up Part B: Observation of a Burning Candle

1. Did you detect any odors? If so, describe them.

_____

2. What color was the molten wax?

_____

3. Did you note the time and date of your candle observations?

_____

4. Was it hotter 6 inches above the flame or 6 inches to the side of the flame (or was it equally hot in both places)?

_____

5. Describe the patterns in the smoke before and after the flame was extinguished.

_____

6. Was the color transition of blue to yellow in the flame gradual or abrupt?

_____

7. Did you include a sketch of the candle in your observations?

_____

CONCEPTUAL PHYSICS	Activity

Making Cents

Purpose
To investigate the relationship between the mass of a penny and its age

Apparatus
10 pennies per student
electronic balance (0.1-gram resolution acceptable; 0.01-gram resolution preferred)
graph paper

Discussion
The scientific method is an effective way of gaining, organizing, and applying new knowledge. A common form of the method is essentially as follows:

1. Recognize a problem.
2. Make an educated guess—a *hypothesis*.
3. Predict the consequences of the hypothesis.
4. Perform experiments to test predictions. If necessary, modify the hypothesis in light of experimental results. Perform more experiments.
5. Formulate the simplest general rule that organizes the three main ingredients—hypothesis, prediction, and experimental outcome.

Procedure
Step 1: Propose a hypothesis to the following question (problem): What effect does aging have on a penny's mass?

_____

_____

Step 2: Based on your hypothesis, predict the general form of a graph that plots the mass of a penny (*y*-coordinate) relative to its age (*x*-coordinate).

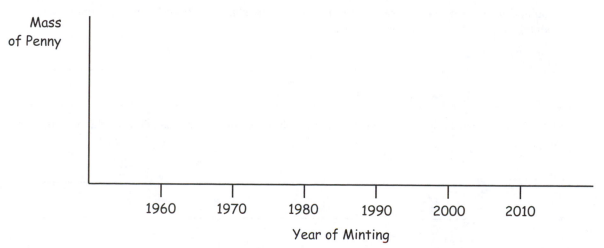

Step 3: Using a balance, measure the mass of at least 10 individual pennies minted in different years. Enter the mass in grams of each penny relative to the year it was minted on the data table.

Data Table

Mass of Penny:										
Year Minted:										

Step 4: Pool your data with that of all other students on the class chalkboard, and create your own graph using all this data. Show the mass of each penny in grams on the *y*-coordinate and the year the penny was minted on the *x*-coordinate. Alternatively, data may be entered into a computer program that will plot the graph for you.

Step 5: Are the experimental results consistent with your hypothesis? If not, propose a new hypothesis.

_____

_____

Step 6: If you have formed a new hypothesis, what additional measurements might you take to support this new hypothesis? Perform these measurements and record your results and observations here:

_____

_____

Summing Up

1. What conclusion can you draw from the results of your experimental data?

_____

_____

2. What effect might aging have on the mass of a nickel, a dime, and a quarter?

_____

_____

3. Would using a balance that was many times more sensitive have made a difference in your conclusion about the effect of aging on a penny? Briefly explain.

_____

_____

4. What improvements might you expect in your graph if only one student had done all the weighing on a single balance?

_____

_____

CONCEPTUAL PHYSICS	Experiment

Chapter 2: Newton's First Law of Motion—Inertia   The Equilibrium Rule

Walking the Plank

Purpose
To measure and interpret the forces acting on an object in equilibrium

Apparatus
meterstick	2 table clamps
2 support rods	2 crossbars (short rods)
2 rod clamps	2 collar hooks
2 spring scales (5- or 10-newton capacity)	slotted masses (two 200 g and one 500 g)
2 20-cm lengths of string	small spirit level (optional)

Discussion
Consider sign painters Burl and Paul who work on a scaffold (a plank of wood suspended by ropes at both ends). They might wonder about the tension in the ropes that support their plank. They are in a state of equilibrium, but how do the rope tensions relate to their weights and the weight of the scaffold? While the weights of Burl and Paul don't change, the tensions in the ropes  do change when either of them moves along the plank. In this activity, you'll use a meterstick for such a scaffold. You'll measure the forces acting on the scaffold when it is in various arrangements, and interpret the forces that determine the condition of equilibrium.

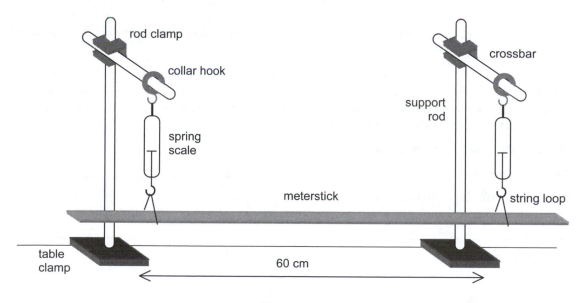

Figure 1

Procedure

Step 1: Calibrate both spring scales so that when held vertically and carrying no load, each reads zero.

Step 2: Arrange the apparatus as shown in Figure 1.

a. Position the table clamps so that the support rods are about 60 cm apart.

b. Attach the crossbars to the support rods using the clamps. Hang a spring scale from each of the crossbars using the collar hooks.

c. Tie the ends of one 20-cm length of string together to create a loop. Hang the loop from one of the spring scales. Repeat for the other spring scale.

d. Suspend the meterstick (centimeter scale up) from the loops of string. Balance the arrangement so that the 50-cm mark is centered between the string loops and the meterstick is level. (Use a spirit level—if one is available—to check the meterstick.) This structure is a model of our painters' scaffold.

e. Adjust the meterstick so that the readings on both spring scales are the same (or very nearly the same). Move the meterstick left or right or adjust the level if necessary.

Step 3: Record the readings on both scales.

Reading on left scale:_____   Reading on right scale:_____

a. Add those readings and record the result. This is the total weight of the meterstick and string loops.

_____

b. Complete the diagram of the meterstick with the forces acting on it. The force L is the upward force on the left, R is the force on the right, and W is the downward force of weight.

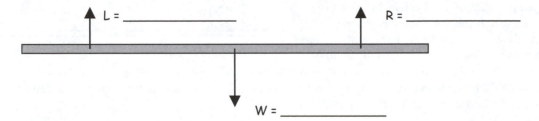

c. What is the net force on the meterstick? The net force is the sum of the forces, taking direction into account.

_____

Step 4: Carefully place one 200-g mass at the 40-cm mark while carefully placing the other 200-g mass at the 60-cm mark. (These represent our painters; take care so they don't fall!) Aim the slots of the slotted masses toward either end of the meterstick (0 or 100 cm).

Step 5: Record the readings for both scales.

Reading of left scale:_____   Reading of right scale:_____

a. What is the total weight of the meterstick, string loops, and masses?

_____

b. Sketch a diagram of the meterstick with all the forces acting on it. Include the numerical values of each force in your diagram.

c. What is the net force on the meterstick?

_____

Step 6: Move the mass at the 40-cm to the 70-cm mark. Keep the other mass at the 60-cm mark. The scaffold is still in equilibrium, even though the load is not evenly distributed.

Step 7: Record the readings for both scales.

Reading of left scale:_____ Reading of right scale:_____

a. What is the total weight of the meterstick, string loops, and masses?

_____

b. Sketch a diagram of the meterstick with all the forces acting on it. Include the numerical values of each force in your diagram.

c. What is the net force on the meterstick?

_____

d. Review your findings so far. What would you say is the condition for equilibrium, a condition that was met in all the arrangements investigated so far?

_____

_____

_____

_____

_____

Step 8: Suppose two painters with different weights used a scaffold. Simulate this by using a 500-g mass and a 200-g mass. Carefully stack the two masses at the 50-cm mark and read the scales.

Reading of left scale:_____   Reading of right scale:_____

Step 9: Carefully place the 200-g mass at the 60-cm mark and the 500-g mass at the 40-cm mark, but do not read the scales yet.

What will the scale readings add to?

_____

_____

Step 10: Read *only* the left scale and record the reading.

Reading on left scale:_____

Predict the approximate value of the reading on the right scale and record your prediction.

Prediction on right scale:_____

Step 11: Read the right scale and record the reading.

Reading on right scale:_____

How did the reading compare to your prediction?

_____

Step 12: Move the 200-g mass until both spring scales have the same reading. Record the location of the 200-g mass.

Position of the 200-g mass:_____

The 500-g mass is 10 cm from the center of the meterstick. How far is the 200-g mass from the center of the meterstick?

_____

Summing Up

1. Can the meterstick platform be in equilibrium if the two upward support forces are equal to each other? If so, give an example from your observations.

_____

_____

_____

2. Can the meterstick platform be in equilibrium if the two upward support forces are unequal? If so, give an example from your observations.

_____

_____

_____

3. Would the platform be in equilibrium if a 500-g mass were at the 30-cm mark and a 200-g mass were at the 60-cm mark? Explain.

_____

_____

_____

4. Suppose the 500-g mass were placed at the 30-cm mark. Where could you place the 200-g mass so that both spring scales would have the same reading? Explain your answer.

_____

_____

_____

5. Could you use the same masses to get both scales to have the same reading if the 500-g mass were placed at the 20-cm mark? If so, where should the 200-g mass be placed? If not, why not?

_____

_____

_____

This experiment centers on an experience with sign painters Paul Hewitt and Burl Grey (see page 25 in the textbook), which led to Paul studying physics. Thanks to Paul's friend, Howie Brand, for suggesting this experiment.

Walking the Plank

CONCEPTUAL PHYSICS | Experiment

Chapter 3: Linear Motion   The Fundamentals of Graphing Motion

Go! Go! Go!

Purpose
To plot a graph that represents the motion of an object

Apparatus
constant velocity toy car
butcher paper (or nonperforated paper towel, or several sheets of paper taped one after another)
access to tape (adhesive or masking)
stopwatch
meterstick
graph paper

Discussion
Sometimes the relationship between two quantities is easy to see. And sometimes the relationship is harder to see. A graph of the two quantities often reveals the nature of the relationship. In this experiment, you will plot a graph that represents the motion of a real object.

Procedure
You will observe the motion of the toy car. By keeping track of its position relative to time, you can make a graph that represents its motion. To do this, you will let the car run along a length of butcher paper. At 1-second intervals, you will mark the position of the car. This will result in several ordered pairs of data—positions at corresponding times. You can then plot these ordered pairs to make a graph representing the motion of the car.

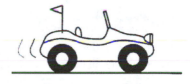

Step 1: Fasten the butcher paper to the top of your table with tape. It should be as flat as possible—no hills or ripples.

Step 2: If the speed of the toy car is adjustable, set it to the "slow" setting.

Step 3: Aim the car so that it will run the length of your table. Turn it on, and give it a few trial runs to check the alignment.

Step 4: Practice using the stopwatch. For this experiment, the stopwatch operator needs to call out something like, "Go!" at each 1-second interval. Try it to get a sense of the 1-second rhythm.

Step 5: Practice the task.

a.   Aim the car to drive down the length of the butcher paper and let it go.

b.   *After* it starts, the stopwatch operator will start the stopwatch and say, "Go!"

c.   Another person in the group should practice marking the location of the front or back of the car on the butcher paper every time the watch operator says, "Go!" For the practice run, simply touch the eraser of the pencil to the butcher paper at the appropriate points.

d.   The watch operator continues to call out, "Go!" (*not* "1, 2, 3 . . . ") once each second, and the marker continues to practice marking the location of the car until the car reaches the end of the butcher paper or table. Take care to keep the car from running off the table!

Step 6: Perform the task.

a. Aim the car to drive down the length of the butcher paper and *let it go*. At this point, *no* marks have been made on the butcher paper. None! The car is moving and no marks have been made.

b. *After* the car begins moving, the stopwatch operator will start the stopwatch and say, "Go!"

c. Another person in the group will mark the location of the front or back of the car on the butcher paper every time the watch operator says, "Go!" **No marks are to be made on the paper until the car is moving.** *Resist your urge to mark the location of the car when it's at rest!*

d. The watch operator continues to call out, "Go!" (*not* "1, 2, 3 . . . ") once each second, and the marker continues to mark the location of the car until the car reaches the end of the butcher paper or table. Take care to keep the car from running off the table!

Step 7: Label the marked points. The first mark is labeled "0," the second is labeled "1," the third is "2," and so on. These labels represent the times at which the marks were made.

Step 8: Measure the distances—in centimeters—of each point from the point labeled "0." (The "0" point is 0 cm from itself.) Record the distances on the data table. Don't worry if you don't have as many data points as there are spaces available on the data table.

Data Table

Time t (s)	0	1	2	3	4									
Position x (cm)	0													

Step 9: Make a plot of position vs. time on the graph paper. Title the graph "Position vs. Time." Make the horizontal axis time and the vertical axis position. Label the horizontal axis with the quantity's symbol and the units of measure: "t (s)". Label the vertical axis in a similar manner. Make a scale on both axes starting at 0 and extending far enough so that all your data will fit within the graph. Don't necessarily make each square equal to 1 second or 1 centimeter. Make the scale so the data will fill the maximum area of the graph.

We could just as easily make a graph of time vs. position. But we prefer position vs. time for a few reasons. In this experiment, time is what we call an "independent variable." That is, no matter how fast or slow our car was, we always marked its position at equal time intervals. *We* were in charge of the time intervals; the *car* was "in charge" of the change in position it made in each interval. But the change in position of the car in each interval depended on the time interval we chose. So we call position the "dependent variable." We generally arrange a graph so that the horizontal axis represents the independent variable and the vertical axis represents the dependent variable. Also, the slope of a position vs. time graph tells us more than the slope of a time vs. position, as we will see later.

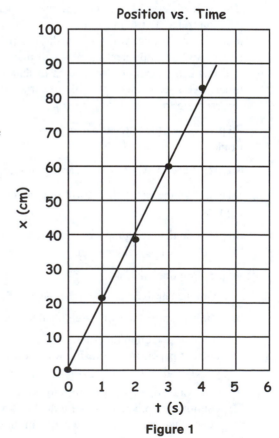

Figure 1

Step 10: Draw a line of best fit. In this case, the line of best fit should be a single, straight line. Use a ruler or straight edge; place it across your data points so that your line will pass as closely as possible to all your data points. The line may pass above some points and below others. Don't simply draw a line connecting the first point to the last point. An example is shown in Figure 1.

Step 11: Determine the slope of the line. *Slope* is often referred to as "rise over run." To determine the slope of your line, proceed as follows.

a. Pick two convenient points on your line. They should be pretty far from each other. Convenient points are those that intersect grid lines on the graph paper.

b. Extend a horizontal line to the right of the lower convenient point, and extend a vertical line downward from the upper convenient point until you have a triangle as shown in Figure 2. It will be a right triangle, because the horizontal and vertical lines meet at a right angle.

c. Find the length of the horizontal line on your graph. This is the "run." Don't use a ruler; the length must be expressed in units of the quantity on the horizontal axis—in this case, seconds of time.

Run: _____ s

d. Measure the length of the vertical line. This is the "rise." Don't use a ruler; the length must be expressed in units of the quantity on the vertical axis. In this case, centimeters of distance.

Rise: _____ cm

e. Calculate the slope by dividing the rise by the run. *Show your calculation* (and include appropriate units) in the space below.

Slope: _____ cm/s

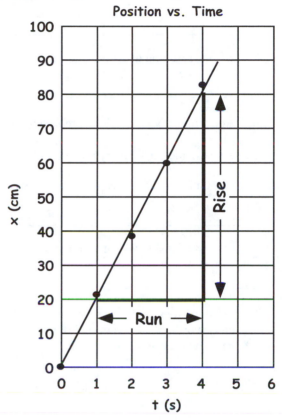

Figure 2

Summing Up

1. Suppose a faster car were used in this experiment. What would have been different about
   a. the distance between the marks on the butcher paper?

   _____

   b. the number of seconds the car would have spent on the butcher paper before reaching the edge?

   _____

   c. the resulting distance vs. time graph? (How would the slope have been different?)

   _____

2. Add a line to your graph that represents a faster car. Label it appropriately.

Go! Go! Go!

3. Suppose a slower car were used in this experiment. What would have been different about

    a. the distance between the marks on the butcher paper?

_____

    b. the number of seconds the car would have spent on the butcher paper before reaching the edge?

_____

    c. the resulting distance vs. time graph? (How would the slope have been different?)

_____

4. Add a line to your graph that represents a slower car. Label it appropriately.

5. Suppose the car's battery ran out during the run so that the car slowly came to a stop.

    a. What would happen to the space between marks as the car slowed down?

_____

    b. Add a line to your graph that represents a car slowing down. Label it appropriately.

_____

6. What motions do these graphs represent? In other words, what was the car doing to generate these motion graphs?

Line A.

_____

_____

_____

_____

Line B.

_____

_____

_____

_____

Position vs. Time

CONCEPTUAL PHYSICS | Tech Lab

Sonic Ranger

Purpose
To produce and interpret real-time graphs of your position using a motion sensor

Apparatus
motion sensor (sonic ranging device)
interface device and connectors
motion sensor software
computer

Discussion
Graphs can be used to represent motion. For example, if you track the position of an object as time goes by, you can make a plot of position vs. time. In this activity, the sonic ranger will track your position, and the computer will draw a position vs. time graph of your motion. The sonic ranger sends out a pulse of high-frequency sound and then listens for the echo. By keeping track of how much time goes by between each pulse and corresponding echo, the ranger determines how far you are from it. (Bats use this technique to navigate in the dark.) By continually sending pulses and listening for echoes, the sonic ranger tracks your position over a period of time. This information is fed to the computer, and the software generates a position vs. time graph.

Procedure
Your instructor will provide a computer with a sonic ranging program installed. Check to see that the sonic ranger is properly connected and operating reliably. Position the sonic ranger so that its beam is about chest high and aimed horizontally. (Note: Sometimes these devices do not operate reliably on top of computer monitors.)

The sonic ranger should be set to "long range" mode. The computer should be set to "graph position vs. time." Initiate the sonic ranger and note how close and how far you can get before the readings become unreliable.

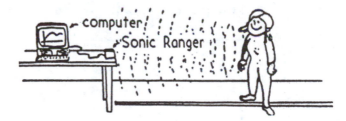

PART A: MOVE TO MATCH THE GRAPH
Generate real-time graphs of each motion depicted on the following pages and write a description of each. *Do not use any form of the term "acceleration" in any of your descriptions.* Instead, use terms and phrases such as "rest," "constant speed," "speed up," "slow down," "toward the sensor," and "away from the sensor."

Study each graph below. When you are ready, initiate the sonic ranger and move so that your motion generates a similar graph. Then describe the motion in words.

Example:

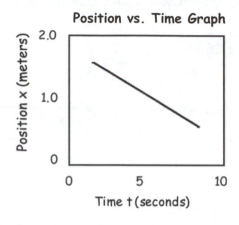

Position vs. Time Graph

Description

Move toward the sensor at constant speed.

Make sure each person in the group can move to match this graph before moving on to the next graph

1.

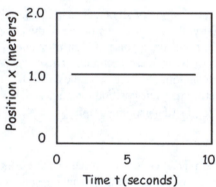

Position vs. Time Graph

Description

2.

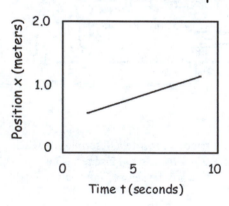

Position vs. Time Graph

Description

3.

Position vs. Time Graph

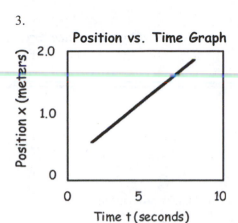

Description

4.

Position vs. Time Graph

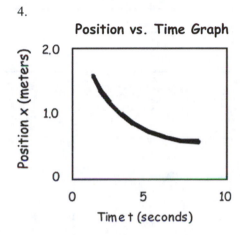

Description

PART B: MOVE TO MATCH THE WORDS

Walk to match each description of motion. Draw the resulting position vs. time graph.

5.

Description

Move toward the sensor at constant speed, stop and remain still for a second, then walk away from the sensor with constant speed.

Position vs. Time Graph

6.

Description

Move toward the sensor with decreasing speed, then just as you come to rest, move away from the detector with increasing speed.

Position vs. Time Graph

Sonic Ranger

7.

Description

Move away from the sensor with decreasing speed until you come to a stop. Then move toward the sensor with decreasing speed until you come to a stop.

Summing Up

1. How does the graph show the difference between forward motion and backward motion?

_____

2. How does the graph show the difference between slow motion and fast motion?

_____

3. Study the graph of position vs. time shown below.

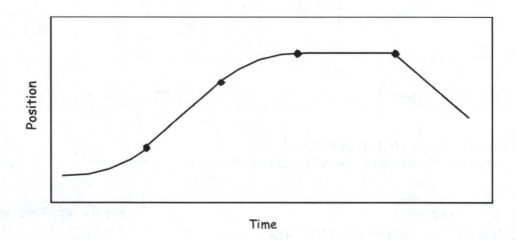

Time

Label sections of the graph showing where the object is
- at rest
- moving forward at constant speed
- moving backward at constant speed
- speeding up
- slowing down

Going Further

Some motion graphing software includes an option for "graph matching." The software displays a plot on the screen, and you must move so that your plot matches the graph on the screen. At the end of the run, the score of your match is displayed. Explore this option and see how high you can score.

CONCEPTUAL PHYSICS

Tech Lab

Motivating the Moving Man

Purpose
To study 1-dimensional motion through the use of a computer simulation

Apparatus
computer   PhET simulation: "Moving Man" (available at http://phet.colorado.edu)

Setup
1. Turn the computer on, log on, and allow it to complete its start-up cycle.

2. Find and start the PhET simulations and run the Moving Man simulation. Ask your instructor for assistance if needed.

3. Locate the time-scale expansion button. Time is the horizontal axis, not the vertical axis. *Use the time-scale expansion button to expand the time axis so that only 10 seconds can be seen.*

4. Find and use the buttons that will delete the velocity vs. time and acceleration vs. time graphs. You are left with one graph: position vs. time. And its horizontal axis now runs from only 0 to 10 seconds.

5. Set the computer aside and read the Discussion section below. When you have finished reading the Discussion, continue with the activity.

Discussion
The simulation provides a simple interface you can use to study motion graphs. Specifically, you can control the position, velocity, and acceleration of an object and see the graphs of position, velocity, and acceleration of that object.

The simulation allows you to manually control the object by dragging it. You can also drag a position "slider" control.

More importantly, you can set initial conditions for the object, then let its motion proceed. You can pause the motion, but the motion is limited in space by the walls.

Procedure
Step 1: Complete the matching exercise on the next page. The names of many objects and buttons are listed. Draw a line connecting the name of each object or button to the object or button itself. Your screen and Figure 1 should be very similar in that only the position vs. time graph is visible and the horizontal (time) scale runs from 0 to 10 seconds. PhET simulations are revised from time to time, so some elements of the screen layout may have changed.

Step 2: Manual Man
a. Click on the man and drag him in the positive direction (toward the house) at a steady rate of 1 m/s to the best of your ability. Your graph won't be perfect (limitations of the computer's interface prevent this), but try to move the man as smoothly as you can at 1 m/s.

b. While the graph is still on the screen, click the on-screen button to show the velocity vs. time graph. This graph plots the value of the velocity as time passes. Clear the graphs and try again. Move the man manually from 0 to 10 meters in 10 seconds—as smoothly as possible—at 1 m/s. The simulation's "stopwatch" reading indicates time.

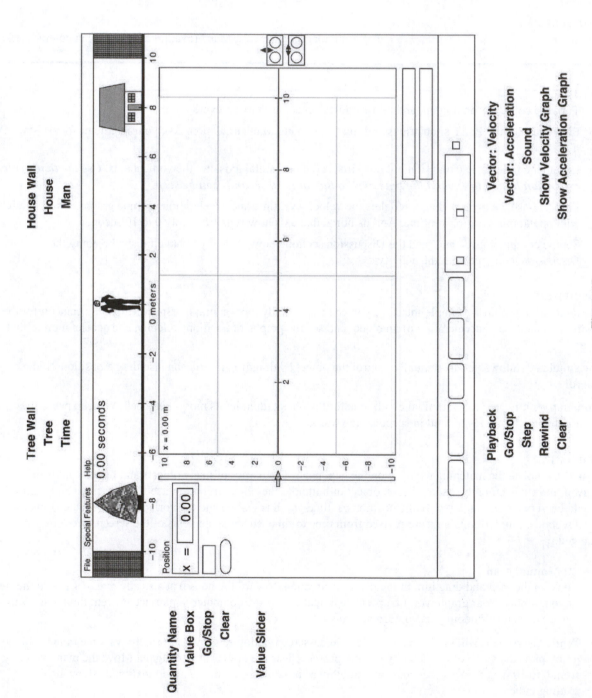

Adjust Vertical Scale
Adjust Horizontal Scale

Tree Wall
Tree
Time

House Wall
House
Man

Quantity Name
Value Box
Go/Stop
Clear

Value Slider

Playback
Go/Stop
Step
Rewind
Clear

Vector: Velocity
Vector: Acceleration
Sound
Show Velocity Graph
Show Acceleration Graph

Figure 1

c. Sketch the resulting position vs. time and velocity vs. time graphs on the graphs page at the end of this lab.

Step 3: Programmed Man
a. Make sure the position and velocity graphs are showing (with the time scale set from 0 to 10 seconds) and the acceleration graph is hidden (deleted). Clear the graphs.

b. Set the initial position of the man to 0 m. (Use the slider or type it into the value box.)

c. Set the initial velocity of the man to +1.0 m/s. (Use the slider or type it into the value box.)

d. Click an on-screen "Go" button and observe the resulting motion graphs. Sketch both graphs (position vs. time and velocity vs. time) on the graphs page at the end of the lab.

e. Clear the graphs. Set the initial velocity of the man to +2.0 m/s and run the simulation again. The motion ends when the man hits the wall. Stop the simulation when that occurs. Disregard data from the wall impact and beyond. Describe *two* differences in the position vs. time graph. (*Hints:* rise/run; duration.)

_____

_____

f. Add lines to your previous sketch representing the +2.0 m/s motion. Be sure to *label* both lines now plotted. And do not include plotted data after the impact with the wall.

g. Add and label a dashed line to the graphs showing the result if the initial velocity of the man were +5 m/s. (Note: Do *not* carry this procedure out on the simulation.)

h. Describe *two* changes that occur on the velocity vs. time graph each time you make the initial speed of the man greater.

_____

_____

Step 4: Back Up the Choo-Choo
a. Clear the graphs. The acceleration graph is still hidden.

b. Set the initial position of the man to 0 m. Set the initial velocity of the man to –2.0 m/s.

c. Run the simulation and observe the motion graphs. Again, the motion ends when Moving Man crashes into the wall, so stop the simulation when that happens. Sketch the graphs on the same set of axes used for Step 3, disregarding data from the wall impact and beyond.

d. Clear the graphs. Set the initial velocity of the man to –4.0 m/s and run the simulation. Describe *two* differences in the position vs. time graph, and add a line to your previous sketch representing the –4.0 m/s motion. Be sure to label both lines now plotted.

_____

_____

e. Add and label a dashed line to the graphs showing the result if the initial velocity of the man were –10 m/s. (Note: Do *not* carry this procedure out in the simulation.)

f. Describe *two* changes that occur on the velocity vs. time graph each time you make the initial speed of the man greater (faster in the negative direction).

_____

_____

g. Click the on-screen button to show the acceleration vs. time graph. What does the acceleration vs. time graph tell you about the motions observed so far?

_____

_____

Step 5: Pickin' Up the Pace

a. Clear the graphs. All three graphs are now showing.

b. Set the initial position of the man to 0 m. Set the initial velocity of the man to 0 m/s. Set the acceleration of the man to +0.5 m/s^2.

c. Run the simulation and observe the motion graphs. Sketch the graphs on the graph page.

d. What does the acceleration vs. time graph tell you about the nature of this motion? (Do not refer to *numerical* values in your response.)

_____

_____

Step 6: Once More, with Feeling

a. Set the initial position of the man to 0 m. Set the initial velocity of the man to 0 m/s. Set the acceleration to the man to +2.0 m/s^2.

b. Run the simulation and observe the position vs. time graph.

    i.   Sketch the graph on the same set of axes as Step 5.

    ii. How is this position vs. time graph different from the one plotted in Step 5 above?

_____

_____

c. How does the velocity vs. time graph differ from the one produced in Step 5 above?

_____

_____

d. Based on what you learned in Steps 5 and 6, sketch all three motion graphs that the man would produce if he started at $x = 0$ m, $v = 0$ m/s, and $a = -1$ m/s^2. Don't actually carry this procedure out on the simulation. Simply sketch what you think it should be, based on your experience.

Step 7: Round Trip

a. With all three graphs showing, adjust the scale of the time axis to display 20 seconds.

b. Set the initial position of the man to –10 m. Set the initial velocity to +3.0 m/s. Set the acceleration to –0.3m/s².

c. Run the simulation and observe the motion graphs. Sketch the graphs on the graphs page.

d. Select the best descriptions of the velocity and acceleration of the man at the *apex* of his motion (when he's farthest from his starting point). Choose one description of the man's velocity and one description of the man's acceleration. Feel free to use the "rewind" and "forward" step features of the simulation.

___The velocity is positive: the man is moving to the right.
___The velocity is zero: the man is at rest.
___The velocity is negative: the man is moving to the left.

___The acceleration is positive: the man's velocity is increasing.
___The acceleration is zero: the man is maintaining constant velocity.
___The acceleration is negative: the man's velocity is decreasing.

e. You overhear a classmate telling someone that it's possible for an object to be at rest *and* accelerating at the same time. What do you think of that statement?

_____

_____

_____

Step 8: The Sloshing Man Puzzle

a. Clear all graphs. Set the position near –2 m. Click and drag the man from –2 m to +2 m with uniform motion and let him rest. Then drag him from +2 m to –2 m with uniform motion and let him rest. Repeat. The resulting position vs. time graph should look similar to the plot below. Don't worry about the time too much; what's important is moving the man one way, then resting, then moving him the other way, then resting. Your lines won't be as straight, and the "corners" of your graph will be rounder than those depicted below.

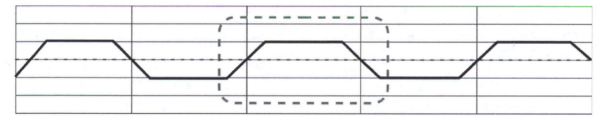

Figure 2. Ideal Position vs. Time for Sloshing Man

b. Examine the section of the plot enclosed in the dashed box. The velocity vs. time graph has a pattern similar to the one shown below. Unlike the time-aligned graphs in the simulation, this graph may have been shifted horizontally. Draw a box to enclose the section of the velocity graph that corresponds to the boxed section of the position vs. time graph above.

Figure 3. Ideal Velocity vs. Time Segment for Sloshing Man

Motivating the Moving Man

c.   The acceleration vs. time graph has a pattern similar to the one shown below. Again, this graph may have been shifted along the time axis. And it is not necessarily to scale. Draw a box to enclose the section of the acceleration graph that corresponds to the boxed section of the position vs. time graph in Figure 2.

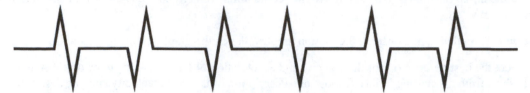

Figure 4. Ideal Acceleration vs. Time Segment for Sloshing Man

d.   Sketch the velocity and acceleration plots in the space below the position plot in Figure 5 below.

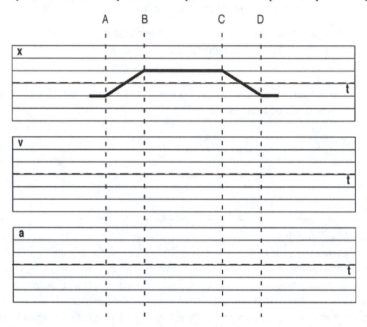

Figure 5. Motion Graph Segment for Sloshing Man

e.   Match the phrases below to the motion illustrated in Figure 5. Use points A, B, C, or D, or segments (e.g., AB, BD).

Sustained positive velocity _____ Sustained negative velocity _____ Rest _____

Positive acceleration to <u>start</u> motion _____ Positive acceleration to <u>stop</u> motion _____

Negative acceleration to <u>start</u> motion _____ Negative acceleration to <u>stop</u> motion _____

Programmed Man and Back Up the Choo Choo

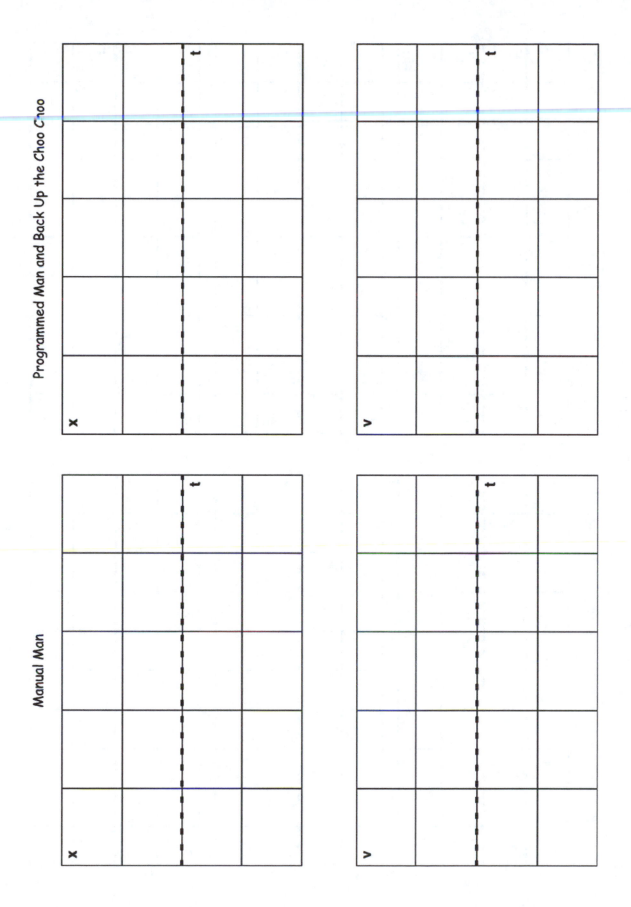

Manual Man

Round Trip

x | | t |
--- | --- | --- | --- |
 | | | |
 | | | |
 | | | |
 | | | |
 | | | |
 | | | |
 | | | |
 | | | |
 | | | |

v | | t |
--- | --- | --- | --- |
 | | | |
 | | | |
 | | | |
 | | | |
 | | | |
 | | | |
 | | | |
 | | | |
 | | | |

a | | t |
--- | --- | --- | --- |
 | | | |
 | | | |
 | | | |
 | | | |
 | | | |
 | | | |
 | | | |
 | | | |
 | | | |

Pickin' Up the Pace and Once More, With Feeling

x | | t |
--- | --- | --- | --- |
 | | | |
 | | | |
 | | | |
 | | | |
 | | | |
 | | | |
 | | | |
 | | | |
 | | | |

v | | t |
--- | --- | --- | --- |
 | | | |
 | | | |
 | | | |
 | | | |
 | | | |
 | | | |
 | | | |
 | | | |
 | | | |

a | | t |
--- | --- | --- | --- |
 | | | |
 | | | |
 | | | |
 | | | |
 | | | |
 | | | |
 | | | |
 | | | |
 | | | |

CONCEPTUAL PHYSICS	Activity

The Weight

Purpose
To investigate the relationship between weight and mass

Apparatus
spring scale (10-newton capacity)
slotted masses (one 100 g, two 200 g, one 500 g)
mass hanger
table clamp
support rod

rod clamp
crossbar (short rod)
collar hook
graph paper

Discussion
Mass and weight are different quantities. Mass is a measure of an object's inertia, the extent to which an object resists changes to its state of motion. Weight is a measure of the interaction between an object and the planet the object is nearest to; usually, that planet is Earth. The weight of an object is related to its mass. In this activity you will find out what that relationship is.

Procedure
Step 1: Check the calibration of the spring scale and adjust it if necessary. The spring scale needs to read zero when there is no load on it. If you're not sure how to do this, ask your instructor how to calibrate the spring scale.

Step 2: Arrange the apparatus as shown in Figure 1.

Step 3: Determine the mass (in grams) and the weight of the mass hanger (in newtons—as shown on the spring scale). Record these values on the second row of the data table.

Step 4: Add 100 grams of slotted mass to the mass hanger. The total mass of the load on the spring scale is now 100 grams plus the mass of the mass hanger. Record the total mass and the weight value (shown on the spring scale) on the next row of the data table.

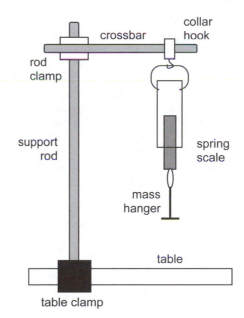

Figure 1

Step 5: Repeat the previous step with 200 g, 300 g, 400 g, 500 g, 600 g, 700 g, and 800 g of slotted mass. Remember that the *total* mass in each case is the sum of the slotted mass and the mass of the mass hanger. When you are done, the column of mass values (in grams) and the corresponding column of weight values (in newtons) will be filled.

Data Table

Total Hanging Mass m (grams)	Total Hanging Mass m (kilograms)	Weight of Mass W (newtons)
0	0	0

Step 6: Convert the mass values to kilograms (1000 grams = 1 kilogram, so 237 grams = 0.237 kg, etc.).

Step 7: Make a plot of weight versus mass, using the mass values in kilograms. As is always the case, the first variable listed in the title of the graph constitutes the vertical axis; the second variable constitutes the horizontal axis. Generally speaking, the first variable listed is the dependent variable (the one you measure during the activity), and the second variable is the independent variable (the one you control during the activity).

Step 8: Title the graph; label the axes to correctly indicate the quantity, units, and scale of each axis.

Step 9: Draw a straight line of best fit through the data points plotted on your graph.

a.   Determine the slope of your best-fit line and record it below.

Slope: _____

b.   What are the units of the slope you found? The slope does have units! Show the units in your answer to the previous question.

The slope of the graph is the relationship between weight and mass. The slope tells how many newtons of weight pull down on each kilogram of mass.

Summing Up

1. What would have been the weight of 1.0 kg? Extend your best-fit line or use a ratio to determine the answer.

_____

_____

2. On the Moon, each kilogram of mass is pulled down with 1.6 newtons of weight. Add a dashed line to your graph showing the results if this activity had been done on the Moon. How does the slope of the **Moon** line compare with the slope of the **Earth** line? Is it steeper (more vertical) or shallower (more horizontal)?

_____

_____

3. If the activity had been done on Jupiter, the resulting line would have had a steeper (more vertical) slope. What does this tell you about the strength of Jupiter's gravitational field as compared with Earth's?

_____

_____

_____

Education is not something that happens to you.

Education is something you do for yourself.

| CONCEPTUAL PHYSICS | Activity |

Putting the Force Before the Cart

Purpose
To observe and interpret the motion of several objects acted on by varying forces to learn the relationship between force, mass, and acceleration

Apparatus

dynamics cart and track	paper clip
mass blocks	4 hex nuts (or equivalent)
string (about 1 meter)	computer with motion graphing software
pulley	motion sensor and interface device
wood block (or end stop)	

Discussion
Some of the most fundamental laws of motion eluded the best minds in science for centuries. One reason for this is that when objects are pushed or pulled, there are usually several forces acting at once. To understand the nature of force and motion, it is necessary to observe the effect of a single, unbalanced force acting on an object. In this activity, you will do just that. You will also vary the amount of force acting on the object. You will then vary the mass of the object being acted upon by a force. Careful observations will lead you to an understanding of the relationship between force, mass, and acceleration—a relationship that Galileo missed, and Newton got!

Procedure
Step 1: Connect the motion sensor to the computer (using the interface device). If the motion sensor has a range selector, choose the "short range" setting. Activate the motion graphing software.

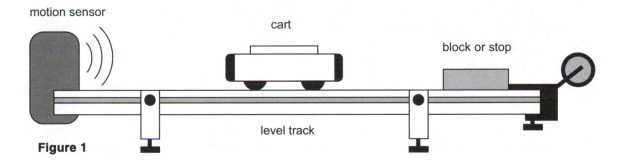

Figure 1

Step 2: Arrange the apparatus as shown in Figure 1.

a. Make sure the track is level. The cart should be able to coast equally in either direction along the track. If the cart "prefers" to roll in one direction, adjust the track accordingly.

b. The pulley clamp should be secure on the track, and the pulley should be able to spin freely.

c. Set the wood block on the track (or employ some other stopping mechanism) so the cart cannot roll into the pulley.

d. Arrange the string so that it is attached to the cart at one end and the paper clip at the other end as shown in Figure 2. The length of the string is such that when the cart is stopped at the wood block, the paper clip does not touch the ground. If the paper clip touches the ground, shorten the string.

Step 3: Test the sensor and software.

a. Place the cart near the middle of the track.

b. Activate the motion sensor and the graphing software.

c. Move the cart back and forth with your hand. The computer should show a graph that corresponds to the motion of the cart. If it does not, adjust the aim of the sensor and check the connecting wires. If the problems persist, ask your instructor for assistance.

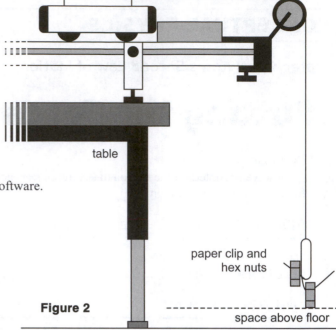

table

paper clip and hex nuts

Figure 2

space above floor

PART A: VARY THE FORCE

Step 4: Check to see that there are two hex nuts attached to the paper clip. Pull the cart back so the paper clip is just below the pulley wheel. Make note of the cart's starting position on the track.

Step 5: Clear the computer of any previous trials and activate the motion sensor.

Step 6: When the motion sensor begins sampling, release the cart and allow it to move along the track until it is stopped by the wood block.

Step 7: Deactivate the motion sensor. If something went wrong during the trial, simply reset the cart, string, and software, and repeat the trial so that you have a reliable result.

a. Sketch the graph in the space to the right. Show the smooth, general pattern; neglect insignificant data point spikes and glitches. Show only the portion of the graph that corresponds to the period **when the cart was moving**. How does this graph of accelerated motion differ from a graph of uniform motion (motion having constant velocity)?

_____

_____

_____

b. What do you think will happen to the acceleration if twice as much force is used to pull the cart?

_____

_____

Step 8: Add the two remaining hex nuts to the paper clip (for a total of four). This will double the force pulling the cart.

Step 9: Set the cart in place for a second trial, starting from the same position on the track. Prepare the software to add a second trial to the one already recorded. Activate the sensor. When the sensor begins sampling, release the cart. When the cart is stopped, deactivate the sensor.

How does the acceleration caused by the doubled force compare with the original acceleration (from Step 6)? Did your observation confirm or contradict your prediction?

PART B: VARY THE MASS

Step 10: Determine the mass of your cart and record it here: _____

Step 11: Clear the computer of any previous trials.

Step 12: Attach four hex nuts to the paper clip. Set the cart in place. Activate the sensor. When the sensor begins sampling, release the cart. When the cart is stopped, deactivate the sensor.

Step 13: Add a mass block or blocks to the cart so that the mass is doubled. For example, if the cart has a mass of 500 g, add 500 g of mass blocks to it. Do not change the hex nut configuration.

What do you think will happen to the acceleration if the same force is used to pull a cart having twice as much mass?

Step 14: Set the cart in place for a second trial. Prepare the software to add a second trial to the one already recorded. Activate the sensor. When the clicking begins, release the cart. When the cart is stopped, deactivate the sensor.

How does the acceleration of the doubled mass compare with the original acceleration (from Step 12)? Did your observation confirm or contradict your prediction?

Putting the Force Before the Cart

Summing Up

1. How does the acceleration of the cart depend on the force pulling it?

   ____ Greater force results in greater acceleration. In other words, acceleration is directly proportional to force.

   ____ Greater force results in lesser acceleration. In other words, acceleration is inversely proportional to force.

   ____ Greater force results in the same acceleration. In other words, acceleration is independent of force.

2. How does the acceleration of the cart depend on the mass of the cart?

   ____ Greater cart mass results in greater acceleration. In other words, acceleration is directly proportional to mass.

   ____ Greater cart mass results in lesser acceleration. In other words, acceleration is inversely proportional to mass.

   ____ Greater cart mass results in the same acceleration. In other words, acceleration is independent of mass.

3. Complete the statement:

   The acceleration of an object is _____ proportional to the net force acting on it and _____ proportional to the mass of the object.

4. Which mathematical expression is most consistent with your observations?

   a. $a = F \cdot m$ \qquad\qquad b. $a = F / m$ \qquad\qquad c. $a = m / F$

5. Examine the position vs. time graphs plotted in the diagram to the right. Suppose plot B represented an empty cart pulled by two hex nuts. If the mass of the cart were doubled and four hex nuts were used to pull the cart, which plot would best represent the result? Explain.

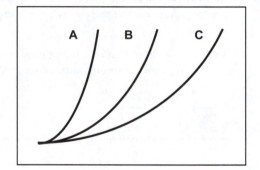

_____

_____

_____

_____

CONCEPTUAL PHYSICS	Activity

Reaction Time

Purpose
To measure your personal reaction time

Apparatus
dollar bill
centimeter ruler

Discussion
Reaction time is the time interval between receiving a signal and acting on it—for example, the time between a tap on the knee and the resulting jerk of the leg. Reaction time often affects the making of measurements. Consider using a stopwatch to measure the time for a 100-meter dash. The watch is started after the gun sounds and is stopped after the tape is broken. Both actions involve reaction time.

Procedure
Step 1: Hold a dollar bill so that the midpoint hangs between your partner's fingers. Challenge your partner to catch it by snapping his or her fingers shut when you release it.

The distance the bill will fall is found using

$$d = \tfrac{1}{2}at^2$$

Simple rearrangement gives the time of fall in seconds.

$$t^2 = \frac{2d}{g}$$
$$t = \sqrt{\tfrac{2}{980}}\sqrt{d}$$
$$t = 0.045\sqrt{d}$$

(For d in centimeters and t in seconds, use $g = 980$ cm/s^2.)

Step 2: Have your partner similarly drop a centimeter ruler between your fingers. Catch it and note the number of centimeters that passed during your reaction time. Then calculate your reaction time using the formula

$$t = 0.045\sqrt{d}$$

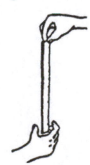

where d is the distance in centimeters.

Reaction time = _____

Summing Up

1. What is your evidence for believing or disbelieving that your reaction time is always the same? Is your reaction time different for different stimuli?

_____

_____

2. Suggest possible explanations why reaction times are different for different people.

_____

_____

3. When might reaction time significantly affect measurements you might make using instruments for this course? How could you minimize its role?

_____

_____

4. What role does reaction time play in applying the brakes to a car in an emergency situation? Estimate the distance a car travels at 100 km/h during your reaction time in braking.

_____

_____

5. Give examples in which reaction time is important in sports.

_____

_____

CONCEPTUAL PHYSICS	**Demonstration**

The Newtonian Shot

Purpose
To explore the role of Newton's laws and gravity as they determine the outcome of a race between two darts fired from two dart guns

Apparatus
2 spring-loaded dart guns that shoot hard-stick, suction-cup darts
2 darts (for the dart guns)
metal ball (steel or lead, approximately 1-inch in diameter)

Discussion
Two darts are to be fired at the same time from two identical dart guns. Both will be fired from the same height and directed straight down toward the ground. One dart has been modified; it is attached to a heavy metal ball.

Procedure
Step 1: The Possibilities. Which ball will hit the ground first? Before you decide, list three possible outcomes and write arguments for each of them. Doing so means you will have to write arguments for two outcomes you don't believe.

a.　Outcome 1: _____

　　Supporting argument: _____

　　_____

b.　Outcome 2: _____

　　Supporting argument: _____

　　_____

c.　Outcome 3: _____

　　Supporting argument: _____

　　_____

Step 2: Your Prediction. Which dart will hit the ground first?

_____

Step 3: Observation. Allow your instructor to complete the demonstration by launching the two darts. Add to the illustration on the back of this sheet. Sketch a snapshot of the moment when the "winning" dart hits the ground, showing the location of both darts at that instant.

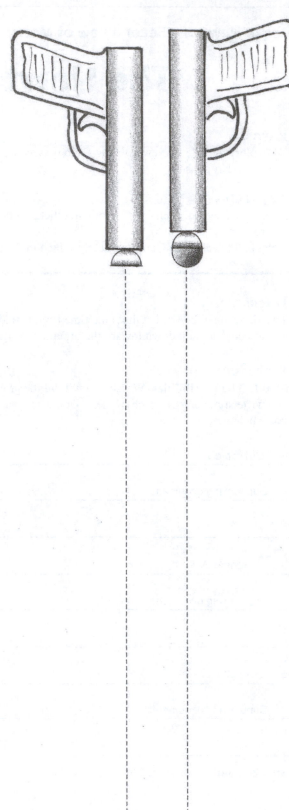

Summing Up

1. Which dart hit first?

_____

2. What is the reason for the outcome observed? Be specific and complete!

_____

_____

_____

_____

_____

| **CONCEPTUAL PHYSICS** | **Tech Lab** |

The Force Mirror

Purpose
To compare the sizes of "action" and "reaction" forces when two objects interact

Apparatus
2 spring scales (preferably with somewhat different ranges)
2 #64 rubber bands (#64 rubber bands are 0.25 inches wide)
computer with software for graphing force sensor data
2 force sensors
interface device(s) for connecting sensors to the computer
table clamp
support rod

Discussion
Forces are interactions between two objects. The interactions can arise in many ways: friction, tension, drag, and many more. But no matter what the nature of the interaction is, two objects are required. There are always two objects, and there are always two forces. The two forces involved in a particular interaction form a *force pair.* You will use spring scales and force sensors to investigate a few force pairs.

Procedure

PART A: SPRING SCALES

Step 1: Attach the table clamp to the table and attach the support rod to the table clamp.

Step 2: Make sure the spring scales are calibrated so that they read 0 when no force is applied to them. If you need assistance, ask your instructor.

Step 3: Take hold of one spring scale while a lab partner holds the other spring scale. Attach a rubber band to the hooks of the spring scales to connect them.

Step 4: While you hold your spring scale in place, have your partner pull his or her spring scale until a low but easy-to-read force is achieved.

Compare your reading to your partner's reading (take care to make sure neither scale is pulled to its full-range limit). Which statement best describes your finding?

____My reading is much greater than my partner's.

____My reading is much less than my partner's.

____My reading is about the same as my partner's.

Step 5: Make a second observation. This time, let your partner hold his or her spring scale in place while you pull your spring scale. How, if at all, does this change the outcome in terms of the force readings?

_____

_____

Step 6: Try other values of force to see if this pattern is consistent. Be careful to make sure you don't reach the limit of either spring scale.

Does the pattern hold across a range of force values?

Step 7: Try connecting one spring scale to the support rod and pulling the other spring scale (while the two are still connected by the rubber band). Does the pattern still hold under these conditions?

PART B: FORCE SENSORS

Step 1: Turn on the computer and let it complete its start-up cycle.

Step 2: Connect the force sensors to the computer using the interface device(s). If you need assistance on this or any of the following steps, ask your instructor.

Step 3: Activate the software that will graph the force sensor data.

Step 4: Use the software to configure the force sensors as follows:
a.   one force sensor should be configured "Push Positive."

b.   the other force sensor should be configured "Pull Positive."

Step 5: Configure the software to show a force vs. time graph of the data from **both** sensors. That is, there will be two plots on one set of axes.

Step 6: Connect the hooks of the force sensors with two rubber bands.

Step 7: Set the force sensors down so that there is no force on their hooks. Press the "zero" button on each sensor to calibrate them.

Step 8: Activate the software's sampling mode (e.g., push the on-screen "Start" button).

Step 9: You and your partner may now pick up the force sensors and pull with varying amounts of force. Continue for about 20 seconds then stop the sampling.

Step 10: Ask your instructor if the resulting graph shows that your sensors are correctly connected and calibrated. If not, make appropriate corrections and try again. Otherwise, continue.

Step 11: Suppose the graph in Figure 1 shows **one** sensor's plot on the force vs. time graph. Based on your experience in this lab, draw the plot of the **other** sensor.

Step 12: Clear the previous force vs. time graph. Attach one force sensor to the support rod. Zero both sensors and connect them with rubber bands. Start sampling and vary the force on the free sensor for about 20 seconds.

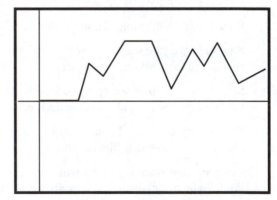

Figure 1. Force vs. Time

Summing Up
Complete the statement. Include the terms "magnitude" (size) and "direction" in your response.

When one object exerts a force on a second object, the second exerts a force on the first that is

CONCEPTUAL PHYSICS	Demonstration

Blowout!

Purpose
To observe a simple blowgun and describe its operation in terms of Newton's three laws of motion

Apparatus
1-inch diameter tube (5- to 10-ft long)
marker pen
masking tape
support rods and clamps (to secure tube)
catch box

Optional Equipment
photogate timing system
balance

Discussion
The operation of a blowgun involves fundamental principles of physics. A force is applied to accelerate a mass to a relatively high speed. The mass travels some distance and is then brought to a stop. Force pairs are involved when the mass is speeding up and slowing down.

Procedure
Step 1: Use the support rods and clamps to hold the tube horizontally. See Figure 1. If the tube is long, or if the tube is PVC, you may need a center support to keep the tube from sagging.

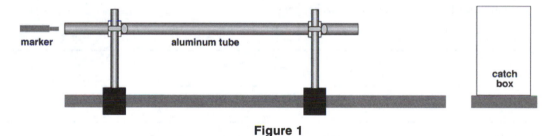

Figure 1

Step 2: Set the catch box some distance from the tube so that it will stop the marker pen when it comes out of the tube.

Step 3: If necessary, wrap the marker pen with enough masking tape so that the taped marker pen barely fits into the tube. See Figure 2. The pen must be free to move through the tube; you'll want a good seal to prevent air from blowing by the marker while it's in the tube.

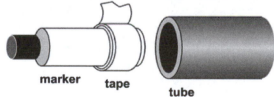

Step 4: Place the marker in the tube and clean the end of the tube so that it is sanitary.

Figure 2

Step 5: Blow the marker out of the tube. Make sure your mouth has a good seal on the tube and blow with as much force as you can.

Summing Up

1. Which of Newton's laws best describes why the marker, initially at rest, requires a force to accelerate it? Why does this law apply to this situation?

_____

_____

2. Identify the force pair involved when the marker is propelled through the tube.

The _____ pushes the _____ forward;

the _____ pushes the _____ backward.

3. a. Which of Newton's laws determines the amount of acceleration that the marker pen experiences while being blown through the tube?

_____

   b. Apply that law to describe the amount of acceleration the marker pen will experience.

_____

_____

_____

4. When the marker pen travels from the tube to the catch box, its motion is best described in terms of Newton's first law. That is to say the marker pen is, for the most part,

___speeding up.   ___slowing down.   ___moving with constant velocity.

5. When the marker pen hits the box, the **marker** pushes the **box** forward.
   a. What **other** force must act when this happens?

_____

_____

   b. This interaction is an example of which law of motion?

_____

Going Further

Step 1: Arrange the photogate system so that it can be used to determine the speed of the marker when it comes out of the tube. See Figure 3.

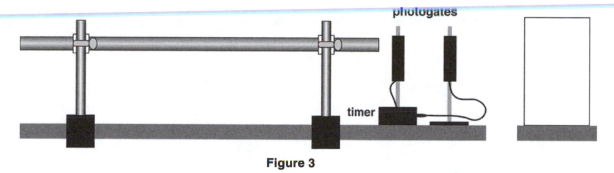

Figure 3

Describe the details of how the speed will be determined.

_____

_____

_____

Step 2: Repeat the demonstration and calculate the speed of the marker pen.

_____

_____

Step 3: Measure the length of the tube. The acceleration of the marker can be determined using the equation $a = v^2/2x$, where v is the speed of the marker and x is the length of the tube. Use the speed of the marker and the length of the tube to calculate the acceleration of the marker pen.

_____

_____

Step 4: Divide the acceleration by 9.8 m/s² to determine the number of g's experienced by the marker pen in the tube.

_____

_____

Step 5: Measure the mass of the marker pen. Use Newton's second law to calculate the force of the exhaled air on the marker pen when the pen is launched.

_____

_____

If a man wants to educate himself, he must first doubt, for in doubting he will find the truth.

Aristotle

CONCEPTUAL PHYSICS	Activity

Egg Toss

Purpose
To investigate the effect that stopping time has on stopping force when momentum changes

Apparatus
raw egg
2 garbage bags
a playing field (soccer, football, baseball, softball, etc.)
access to 2 bright (polypropylene) ropes (optional)
trundle wheel or 100-foot tape measure (optional)

access to masking tape
safety glasses or goggles

Procedure
One person in the group is going to throw a raw egg to another person in the group. The second person must catch the egg without letting it break. When the thrower and catcher are close to each other, the task is fairly simple. As the distance increases, the task becomes more difficult.

Step 1: Prepare for the activity by choosing a thrower and a catcher. The thrower will throw the egg to the catcher. The catcher will catch the egg. Both thrower and catcher must use only their bare hands to handle the egg.

Step 2: The catcher must wear the safety goggles and plastic clothing protection. Use the garbage bags and masking tape to construct plastic ponchos and skirts (or kilts). When the egg breaks, it can make a mess. Make sure the plastic protective garments cover any part of your clothes that should be protected from raw egg white and yolk.

Step 3: Go to the field and line up according to your instructor's directions. The throwers and catchers should be facing each other and should start about 3 meters apart from each other. Throwers should also be about 3 meters from each other.

If possible, line up so that the sun is not shining in the eyes of the throwers or the catchers.

If using the ropes, lay both ropes on the ground and pull them straight to serve as "foot-fault" lines—one in front of the throwers and the other in front of the catchers. Neither throwers nor catchers may step over their respective ropes.

Step 4: When the instructor gives the signal, the thrower throws the egg to the catcher.

If the catcher catches the egg and the egg remains intact, the group may proceed to the next toss.

If the egg is not properly caught but remains intact, the group must repeat the toss.

If the egg breaks, step on the remains to grind them into ground. If a trundle wheel or long tape measure is available, determine the distance between the thrower and catcher when the egg broke.

The thrower (or someone else in the group) must retrieve the egg from the catcher. The catchers should move back three "giant steps" (3 meters), and the throwers should return to their original throwing line. If using ropes, move the catchers' foot-fault line back and stretch it straight.

Step 5: Repeat Step 4 until the last group breaks their egg.

Summing Up

1. What was the maximum distance between thrower and catcher before the egg broke? What was the greatest distance achieved in the class? (Use an estimate if you didn't measure it.)

_____

_____

2. What was the trick to making a successful catch? What does this have to do with stopping time?

_____

_____

3. Compare a sudden-stop catch with a gradual-stop catch.
   a. In which case is the mass of the egg greater? Or is it the same either way?

_____

   b. In which case is the change in velocity of the egg greater? Or is it the same either way? (Be careful!)

_____

   c. In which case is the change in momentum ($m\Delta v$) of the egg greater? Or is it the same either way?

_____

   d. In which case is the stopping time greater? Or is it the same either way?

_____

   e. In which case is the stopping force greater? Or is it the same either way?

_____

4. Use your findings from this activity to explain the purpose of airbags in cars. ***Don't*** use words like "cushion," or "absorb." ***Do*** use terms like "stopping time," and "stopping force."

_____

_____

_____

5. What are some other examples of changing stopping time to change stopping force?

_____

_____

_____

| **CONCEPTUAL PHYSICS** | **Activity** |

Bouncy Board

Purpose
To investigate the effect of stopping time when momentum changes

Apparatus
table
meterstick
short length (30 to 40 cm) of weak string (mailing parcel twine or equivalent)
various masses

Discussion
In bungee jumping, it is important that the cord supporting the jumper stretches. If the cord has no stretch, then when fully extended it either brings the jumper to a sudden halt or the cord snaps. In either case, ouch! Whenever a falling object is brought to a halt, the force that slows the fall depends on the time it acts. You will see evidence of this in this activity.

Procedure
Through the small hole at the end of a meterstick, thread a piece of string about 30 cm long and tie it in place. (If there is no hole, consider drilling one, or tying the string very tightly.) Attach a mass to the other end of the string—try a kilogram for starters. Place the stick on a tabletop and slide the stick so that most of its length extends over the edge of the table, Figure 1. While holding the stick firmly to the table, hold the mass slightly beyond the edge of the stick, then drop the mass. The string and the bending of the stick should stop its fall. The string shouldn't break (if it does break, try a stronger string or a smaller mass—experiment). Now

Figure 1

Figure 2

repeat this activity, but with only a small portion of the stick extending over the table's edge. See Figure 2. What happens now?

Try different configurations to see what conditions result in string breaking. Experiment with different masses, different amounts of meterstick overhang, and different string lengths.

Summing Up

1. Why is it important that a bungee jumper be brought to a halt gradually?

_____

_____

2. How does the *impulse = change in momentum* formula, $Ft = m\Delta v$, apply to this activity?

_____

_____

3. Exactly why did the string break when there was less "give" to the meterstick?

_____

_____

4. The breaking strength of the string most certainly plays a role in this activity. How does the length of the string play a role?

_____

_____

5. How does the falling mass play a role? Will twice the mass require twice the stopping force if it is brought to a halt in the same time? Defend your answer.

_____

_____

6. Why is it important that fishing rods bend?

_____

_____

| **CONCEPTUAL PHYSICS** | **Experiment** |

An Uphill Climb

Purpose
To determine what advantage—if any—there is in using an inclined plane to move an object to a higher elevation

Apparatus
dynamics cart and track or board that can be inclined and secured at various angles
spring scale (capable of weighing the dynamics cart)

table	table clamp
support rod	rod clamp
meterstick	protractor

Discussion
Why are ramps used when lifting heavy objects? Does it make the task easier (requiring less force)? Does it make the movement shorter (requiring less distance)? Does it make the effort more efficient (requiring less work)? Perhaps it does several of these; maybe it does none of them. You will learn more from this lab if you record your initial thoughts before making any measurements or calculations.

What advantages or disadvantages are there in using a ramp when lifting a heavy object?

_____

_____

_____

In this experiment, the cart will act as the heavy object. Your task will be to move your cart a vertical distance of 20 cm above the tabletop. You will arrange a series of ramps (inclined planes) at different angles to accomplish this task. You will measure the force needed to move a cart up the incline. You will also measure the distance through which that force would be applied to finish the job. You will then calculate the work required to lift an object using an inclined plane.

By the end of the experiment, you will be able to identify what an inclined plane can do for you in terms of force, distance, and work.

Procedure

PART A: SHALLOW INCLINE
Step 1: Arrange the apparatus as shown in Figure 1. The plane should be inclined at an angle between 20° and 30°. Check the angle with the protractor as shown in Figure 2.

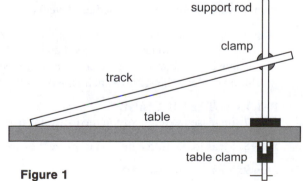

Figure 1

An Uphill Climb

Step 2: Measure the force needed to pull the cart along this incline using the spring scale as shown in Figure 3. The spring scale is held parallel to the inclined plane when the measurement is made. Because the force needed to move the cart at a constant speed is the same as the force needed to keep the cart at rest on the plane, measure the force when the cart is at rest. Record the force below and transfer the value to the data table.

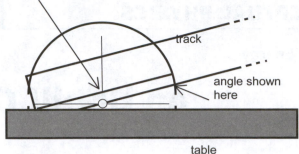

Figure 2

Shallow incline force:

F = _____ N

Step 3: Measure the distance the cart would travel along the inclined path to move from the tabletop to a distance 20 cm above the tabletop. See Figure 4. The path starts at the tabletop (even if the object being used as an incline plane doesn't come all the way down to the tabletop). The path ends where the inclined plane is 20 cm above the tabletop. This distance will be greater than 20 cm for all inclined planes (unless the plane goes straight up). Convert the distance from centimeters to meters. Record the distance below and transfer the value (in meters) to the data table.

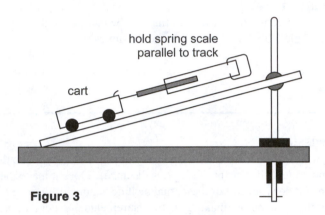

Figure 3

Shallow incline distance:

d = _____ cm

= _____ m

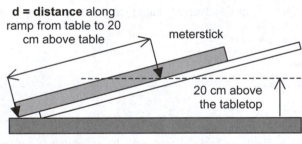

Figure 4

PART B: MEDIUM INCLINE
Step 4: Increase the angle of incline to a value between 40° and 50°.

Step 5: Measure the force needed to move the cart along this incline and record it on the data table.

Step 6: Measure the distance the cart would travel along this path to move 20 cm above the tabletop. The upper end of the incline is much higher now than it was for the shallow incline, but your only concern is moving the cart 20 cm above the tabletop. The distance the cart would travel along this path will be smaller than the distance it would travel along the shallow incline. Record the distance (in meters) on the data table.

PART C: STEEP INCLINE
Step 7: Increase the angle of incline to a value between 60° and 70°.

Step 8: Measure the force needed to move the cart along this incline and record it on the data table.

Step 9: Measure the distance the cart would travel along this path to move 20 cm above the tabletop. Record the distance (in meters) on the data table.

PART D: STRAIGHT UP

Step 10: Measure the force needed to move the cart straight up. No incline is needed for this task. Record the force on the data table.

Step 11: The distance the cart would travel along this path to move 20 cm above the tabletop is 20 cm. Record the distance (in meters) on the data table.

PART E: CALCULATING WORK

Step 12: Calculate the work done along the shallow incline. Multiply the force applied to the cart by the distance the cart traveled to move 20 cm above the tabletop. Show the calculation for the shallow path below and transfer the result to your data table. Include the correct units in all your values.

W = F · d = _____ x _____ = _____

Step 13: Calculate the work for the other paths: the medium incline, steep incline, and straight up. Record the results on the data table.

Step 14: Show your instructor your completed data table before proceeding to the Summing Up section.

Data Table

	Angle θ (degrees)	Force F (newtons)	Distance d (meters)	Work W (joules)
Shallow Path				
Medium Path				
Steep Path				
Straight Up				

An Uphill Climb

Summing Up

1. As the incline gets steeper, what happens to the force required to pull the cart?

    ____The force increases significantly.

    ____The force decreases significantly.

    ____The force remains about the same.

    (Compare the force needed to pull the cart along the shallow path with the force needed to pull the cart straight up. A significant difference is one that is 20% or greater.)

2. As the incline gets steeper, what happens to the distance traveled by the cart?

    ____The distance increases significantly.

    ____The distance decreases significantly.

    ____The distance remains about the same.

    (Compare the distance along the shallow path with the distance of the path straight up.)

3. As the incline gets steeper, what happens to the work required to move the cart 20 cm above the tabletop?

    ____The work increases significantly.

    ____The work decreases significantly.

    ____The work remains about the same.

    (Compare the work needed to move the cart up the shallow path with the work needed to move the cart straight up.)

4. What is the *advantage* of using an inclined plane rather than moving something straight up?

    _____

    _____

    _____

5. What is the *disadvantage* of using an inclined plane rather than moving something straight up?

    _____

    _____

    _____

6. The work done to move something is a measure of the energy required to complete the task. The energy required to move an automobile is provided by the fuel. Would it be more fuel-efficient to drive to the top of a hill along a steeply inclined road or a gradually inclined road? Explain your answer in terms of what you observed in this experiment.

    _____

    _____

    _____

| **CONCEPTUAL PHYSICS** | **Demonstration** |

The Fountain of Fizz

Purpose
To observe a Mentos geyser and measure several quantities involved in the eruption, then determine the speed of the fizz and the power of the eruption.

Apparatus
2-liter bottle of diet soda pop
7 Mentos candies
mechanism for dropping the candies into the soda pop
electronic balance
stopwatch
means of measuring the height of the eruption (which may exceed 5 meters)

Discussion
The Mentos geyser is a popular science demonstration. Several Mentos candies are dropped into an open bottle of diet soda pop, and the liquid rapidly turns to fizz. The fizz shoots out of the bottle to great heights. The chemistry involves the carbon dioxide in the soda pop. The nature of the surface of the Mentos candies and an ingredient (gum arabic) in the candy are the significant factors. They cause the carbon dioxide to come out of solution, creating the fizz. In this activity, you will focus on the *physics* of the demonstration.

Procedure
Step 1: Determine and record the mass of the demonstration ingredients *before* the demonstration. The ingredients are the 2-liter bottle of soda pop and the Mentos (seven Mentos seems to work well).

     Initial Mass = _____

Step 2: Prepare the eruption in an open, outdoor space where splashing diet soda pop will not create a problem.

Step 3: Prepare a means to measure the maximum height of the eruption.

Step 4: Prepare to measure the duration of the eruption. That is, you will start timing when the fizz first emerges from the bottle and stop timing when the column of fizz collapses.

Step 5: When preparations are complete, activate the demonstration by dropping the candies into the soda pop and observe the height and time of the eruption.

     Maximum Height of Eruption = _____

     Time of Eruption = _____

Step 6: Measure and record the mass of the ingredients (Mentos, soda pop, and bottle) that remain after the eruption has concluded.

     Final Mass = _____

Summing Up

1. What type of energy (kinetic or potential) does the fizz have when it emerges from the bottle on its way up?

_____

2. What type of energy does the fizz have when it reaches the top of its flight and is about to come back down?

_____

3. How much mass was ejected from the bottle during the eruption?

_____

_____

4. Assume that all the mass ejected in the eruption rose to the maximum height measured during the demonstration. Calculate the potential energy of all the fizz at that height. (Start by writing the equation for potential energy.)

_____

5. According to the principle of conservation of energy, the potential energy of the fizz at the top of the flight is equal to the kinetic energy of the fizz when it emerges from the bottle. Write the equation for the kinetic energy of an object in terms of its mass and speed.

_____

6. Rearrange the equation to solve for the speed of the soda pop as it emerges from the bottle.

_____

_____

_____

7. Power is the rate at which work is done or energy is transformed. That is,

Power = Energy/Time

Use the energy found above and the time measured during the demonstration to calculate the power developed in the eruption.

_____

_____

_____

CONCEPTUAL PHYSICS　　　　　**Experiment**

Dropping the Ball

Purpose
To determine and compare the potential energy of the ball before it's dropped with the kinetic energy of the ball after right before it hits the ground

Apparatus
acrylic tube (1.0-inch diameter, about 4 feet in length)
steel ball, about 16 mm
small rare-earth magnet (neodymium or equivalent)
2 photogates and timers
table
table clamp
support rod
2 three-finger clamps or buret clamps
meterstick

SAFETY NOTE: Use caution when handling the magnet to avoid pinching. Keep it away from computers, sensitive electronic devices, and magnetic storage media such as computer disks.

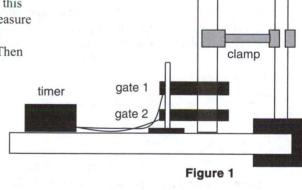

Figure 1

Discussion
Your textbook gives several examples of work done in lifting an object, transforming the work to potential energy, and then transforming it to kinetic energy when the object falls. In this experiment you'll do the same with a steel ball. You'll measure its potential energy when lifted to a certain height. You'll measure its kinetic energy after falling from that height. Then you'll compare the two energy values.

Procedure

Step 1: Arrange the apparatus as shown in Figures 1 and 2. The upper photogate (gate 1) should be about 5 cm above the lower photogate (gate 2). Connect both gates to the timer.

Step 2: Configure the photogate timer to read the time between the two gates (that is, the timer starts when the beam of the first photogate is interrupted and stops when the beam of the second photogate is interrupted).

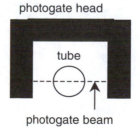

Figure 2. Top View
Make sure the photogate beam passes through the diameter of the tube.

Step 3: Measure the distance between the photogate beams as shown in Figure 3. *Be very careful taking this measurement.* Record the distance in centimeters and convert it to meters.

Distance between photogate beams:

d = _____ cm

= _____ m

Step 4: Determine the mass of the ball. Record its mass in grams and kilograms.

Mass of steel ball:

m = _____ g

= _____ kg

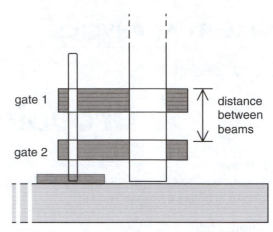

Figure 3

Step 5: Set the ball inside the tube. Use the magnet to lift the ball up through the tube so that the bottom of the ball is 40 cm above the upper photogate as shown in Figure 4. Be careful: The bottom of the ball must be 40 cm above the **upper photogate**, not 40 cm above the **table**.

Step 6: Clear or "arm" the photogate timer so that it is prepared to make a measurement.

Step 7: *Carefully* remove the magnet from the side of the tube. Doing so will release the ball to fall to the bottom of the tube. When the ball passes through the photogate beams, a measurement will be made and displayed on the timer. The goal is to release the ball from rest, so take care not to give the ball upward or downward motion when you release it.

Step 8: If the trial went well, record the time value. Repeat the process until you have three reliable time values. (If you make a mistake during a trial, do not record the result. Simply repeat the process until you have a good trial.)

Time values: _____

_____

_____

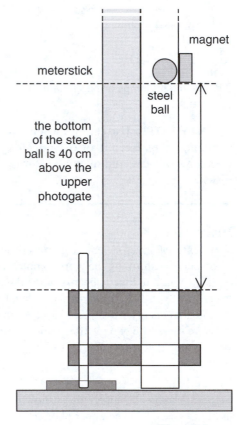

Figure 4

Step 9: Determine the average of the three values and record it in the data table.

Step 10: Repeat the process for drops from 60 cm, 80 cm, and 100 cm. Record the results in the data table.

Data Table

Drop Height h (meters)	Photogate Time t (seconds)	Speed v (m/s)	Potential Energy PE (J)	Kinetic Energy KE (J)
0	—	0	0	0
0.40				
0.60				
0.80				
1.00				

a. Calculate the potential energy of the ball when it was 40 cm above the photogate using the equation $PE = mgh$ (m is the mass of the ball, g is 9.8 m/s^2, and h is 0.40 m). Show your work and your solution. (The value should be between 0.050 J and 0.100 J.) Record your solution in the data table as well.

_____

_____

_____

b. Calculate the potential energy of the ball at the other heights and record your solutions in the data table.

c. Calculate the speed of the ball after it fell 40 cm using the equation $v = \dfrac{d}{t}$ (d is the distance between the photogates and t is the time it took the ball to pass between the photogates). Show your work and your solution below. (The value should be between 2.50 m/s and 3.00 m/s.) Record your solution in the data table.

_____

_____

_____

d. Calculate the speed of the ball after falling from the other drop heights and record your solutions in the data table.

e. Calculate the kinetic energy of the ball after it fell 40 cm using the equation $KE = \dfrac{1}{2}mv^2$ (m is the mass of the ball, and v is the speed of the ball). Show your work and your solution. (The value should be between 0.050 and 0.100 J.) Record your solution in the data table.

_____

_____

_____

f. Calculate the kinetic energy of the ball after falling from the other drop heights, and record your solutions in the data table.

Dropping the Ball

Summing Up

1. When the drop height doubles from 40 cm to 80 cm, which of the following quantities also doubles (approximately)?

    ____ speed after the fall

    ____ potential energy at the drop height

    ____ kinetic energy after the fall

2. Which statement best describes the relationship between the potential energy at the drop height to the kinetic energy after the fall?

    ____ The potential energy is always significantly higher than the kinetic energy.

    ____ The kinetic energy is always significantly higher than the potential energy.

    ____ The potential energy and kinetic energy are about the same.

    (A significant difference in this experiment would be a difference of 20% or more. See Appendix C for information on how to calculate percent difference.)

3. Use your findings to predict the following values for a trial involving dropping the ball from 160 cm.

    a. Potential energy = _____

    b. Kinetic energy = _____

    c. Speed after falling = _____

    Hint: By what factor was the "80-cm ball" faster than the "40-cm ball"? Use this factor to determine how much faster the "160-cm ball" will be compared with the "80-cm ball."

Going further

When the ball was released from a particular height, its potential energy was transformed to kinetic energy as it fell. This type of energy transformation happens on a roller coaster as well. Potential energy that the roller coaster has at the top of the first hill is transformed to kinetic energy as it rolls downward. In an ideal roller coaster (with no frictional losses), the kinetic energy at the bottom of the hill would be equal to the potential energy at the top.

Consider an ideal roller coaster. The first hill has a certain height. When the roller coaster reaches the bottom of the hill, it is traveling at a certain speed.

1. How much higher would the hill have to be so that the roller coaster had twice as much *kinetic energy* at the bottom of the hill?

    _____

2. How much higher would the hill have to be so that the roller coaster had twice as much *speed* at the bottom of the hill? (The answer to this question is not the same as the answer to the previous question.)

    _____

3. In real-world roller coasters, each hill is shorter than the hill before it. Why do you suppose that is?

    _____

    _____

Name _____ Section _____ Date _____

CONCEPTUAL PHYSICS	Activity

A Puzzle—With a Twist

Twin-Baton Paradox

Purpose
To experience rotational inertia "firsthand" by observing the connection between the distribution of mass and resistance to rotational acceleration

Apparatus
1 pair of rotational batons
meterstick
spring scale

Discussion
The extent to which an object maintains its state of motion is its inertia. Mass is a measure of inertia. More massive objects require more force to undergo a given acceleration. The laws of linear motion apply to rotation as well. The extent to which an object maintains its state of rotational motion is its rotational inertia. Objects with more rotational inertia require more torque to undergo a given angular acceleration.

In this activity, you will assess linear and rotational inertia "firsthand" by shaking and twisting a pair of batons. There is something different about these seemingly identical batons.

Procedure
Step 1: Measure and record the length of each baton.

Length of Baton A: _____

Length of Baton B: _____

Is one baton significantly longer than the other or are the batons essentially equal in length?

_____

Step 2: Measure and record the mass of each baton.

Mass of Baton A: _____

Mass of Baton B: _____

Is one baton significantly more massive than the other or are the batons essentially equal in mass?

_____

Step 3: Stand up and hold one baton in each hand. While keeping each baton vertical, shake them left and right (inward and outward) as rapidly as you can. See Figure 1.

Figure 1. Shaking the Batons Back and Forth

The extent to which the batons resist the back and forth motion is an indication of their "linear inertia" (regular inertia). Try it once, then switch the batons so that the hand holding Baton A is now holding Baton B.

Does one baton resist being shaken significantly more than the other, or do the batons have essentially the same linear inertia?

_____

_____

Step 4: Now hold the batons vertically with your arms extended. Instead of shaking the batons back and forth, twist the batons at least 90 degrees left and right as rapidly as you can. See Figure 2.

Figure 2. Twisting the Batons Back and Forth

The extent to which the batons resist the back and forth twisting is an indication of their rotational inertia. Try it once, then switch the batons so that the hand holding Baton A is now holding Baton B.

Does one baton resist being twisted significantly more than the other, or do the batons have essentially the same rotational inertia?

_____

_____

Figure 8.9 in the textbook illustrates how the difficulty in rotating a dumbbell depends on the position of the bells. Rotational inertia depends on mass distribution.

Summing Up

1. One baton has a significant amount of mass located near its center. The other has mass located at the ends. Correctly label the illustrations in Figure 3, "Baton A, " and "Baton B."

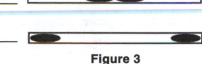

Figure 3

2. Which dumbbell in Figure 4 would have more rotational inertia (resistance to twisting)? Why?

_____

_____

_____

Figure 4

3. Which dumbbell in Figure 5 would have more rotational inertia? Why?

_____

_____

_____

Figure 5

4. ***Figure 6 is not drawn to scale.*** Dumbbell G has two 4-kilogram masses 2 meters apart; Dumbbell H has two 1-kilogram masses 4 meters apart. Dumbbells G and H have ***equal*** rotational inertias. Dumbbells J and K also have equal rotational inertias.

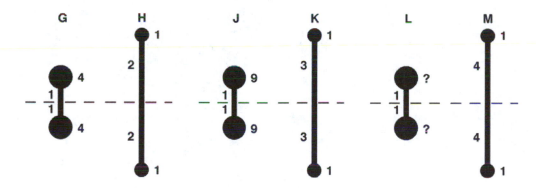

Figure 6

How much mass must each ball on Dumbbell L have so that its rotational inertia is equal to that of Dumbbell M? Dumbbell M has two 1-kilogram masses 8 meters apart.

Twin-Baton Paradox

Doing experiments in the lab is the clincher to understanding physics.

| **CONCEPTUAL PHYSICS** | **Activity** |

It's All in the Wrist

Purpose
To experience torque "firsthand" by observing the dependence of torque on the so-called "lever arm"

Apparatus
meterstick
meterstick clamp
500-gram hooked mass *or* mass hanger and 500-gram slotted mass

Going Further
aluminum pipe (5 ft)
bottled water (half-liter, in a plastic bottle)
string
strong tape

Discussion
Rotational (angular) mechanics has many connections to translational (linear) mechanics. In "Twin-Baton Paradox," you learned about the angular version of mass: ***rotational inertia.*** It's one thing to learn about it by reading the definition and manipulating the equation. It's another thing to learn it through a hands-on experience.

This activity focuses on the angular version of force: ***torque.*** While a force can be described as a push or a pull, torque is a twist. An unbalanced force can cause linear acceleration. An unbalanced torque—a twist—can produce angular acceleration.

Procedure
Step 1: Hold the end of the meterstick so that your index finger is between the 5-cm and 10-cm mark. Position the clamp at the 20-cm mark and suspend the 500-g mass (the load) there. See Figure 1.

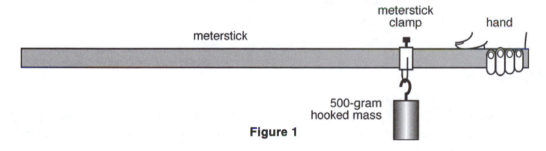

Figure 1

Step 2: Rotate the meterstick slightly (about 15° up and down) so that the free end is raised and lowered while the end you hold does not rise or fall. Do this ***slowly*** and ***carefully***; do not allow the weight to hit anything during this process, because doing so may cause it to fall off the hanger. Note the level of difficulty associated with rotating this configuration.

Step 3: Move the load to the 50-cm mark without moving your hand from the original pivot-point. Rotate the meterstick up and down again.

Step 4: Repeat this procedure with the mass at the 90-cm mark. Keep your hand at the 10-cm mark pivot-point so that the hanging weight is getting farther and farther from your hand.

When a load is applied along the length of a meterstick held at one end, a *torque* is produced. To keep the meterstick-mass assembly from falling to the ground, your wrist must apply an equal torque in the opposite direction.

The "torque feeler" illustrated in Figures 8.16 and 8.17 in the textbook convincingly shows the different effects of force and torque.

Summing Up

1. What happens to the *weight* of the load as it is moved farther from your wrist? (Does it increase, decrease, or remain constant?)

   _____

2. What happens to the *torque* of the load as it is moved farther from your wrist? The distance from the load (F) and the pivot-point (your wrist) is called the *lever arm* (r).

   _____

3. Suppose the load were hanging at the 50-cm mark. What do you suppose would happen to the torque exerted on your hand if the weight of the load were doubled?

   _____

4. a. As the lever arm (r) increases, the torque _____

   b. As the applied force (F) or load increases, the torque _____

5. Which expression of the relationship between torque (τ), lever arm (r), and applied force (F) is most consistent with your experience in this activity?

   ____ $\tau = r \times F$ ____ $\tau = r/F$ ____ $\tau = F/r$

6. Which arrangement, if either, would apply more torque to your wrist?

   ____ a 500-gram mass suspended 30 cm from your wrist

   ____ a 250-gram mass suspended 60 cm from your wrist

   ____ actually, it's the same amount of torque for both arrangements

Going Further

THE TORQUEMASTER CHALLENGE

Use some string to tie a half-liter bottle of water to one end of an aluminum pipe. Secure the string to the pipe with strong tape if necessary. The water bottle should dangle freely while tethered securely to the pipe.

The instructor (or designated spotter) holds the water bottle end of the pipe waist high. The Torque Challenger (contestant) holds the free end waist high. The pipe is now horizontal (more or less). When the signal is given, the spotter releases the water bottle end of the pipe. The Torque Challenger must now keep the pipe horizontal for 10 seconds to become a Torquemaster.

Exceptionally gifted Torquemasters can raise the pipe while keeping it horizontal. They do this by holding the pipe forward as if to strike a fencing posture.

CONCEPTUAL PHYSICS	Demonstration

Will it Go 'Round in Circles?

Purpose
To investigate the nature of acceleration in circular motion—its direction and the factors that determine its magnitude

Apparatus
dynamics cart
dynamics cart track or level table
2 Visual Accelerometers
low-friction rotational platform

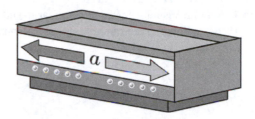

Going Further
Newton's cradle
Laser Viewing Tank or liquid accelerometer

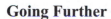

Discussion
By definition, acceleration is the rate at which velocity changes. As simple as that definition may seem, acceleration is one of the most sophisticated concepts you will encounter in this course.

For example, there is acceleration whenever velocity changes. By way of review, list three **_different_** ways that the velocity of an automobile could change, and identify the control mechanism that the driver would use to make each change. (Note: Each method requires a different control mechanism.)

1. _____

2. _____

3. _____

Draw a box around the velocity change that doesn't require a change in the vehicle's speed. That's the type of acceleration you'll be looking at during this demonstration.

Procedure

PART A: THE VISUAL ACCELEROMETER

Step 1: Attach a Visual Accelerometer to a dynamics cart and place the cart on a wide level surface or (preferably) a track. Turn the accelerometer on and set the range for "Manual: 5 m/s^2."

Step 2: Arrange volunteers at opposite ends of the track or level space.

Step 3: Have the volunteers "play catch" with the dynamics cart by gently pushing it back and forth between the ends. The pusher should give the cart a gradual acceleration rather than an abrupt one. The catcher should slow the cart gradually rather than abruptly stopping it.
a.   The volunteer at the left end pushes the cart to the right; the volunteer at the right end catches it.
b.   The volunteer at the right end pushes the cart to the left; the volunteer at the left end catches it.
c.   Repeat several times for careful observation. Look for a connection between the motion of the cart and the LEDs on the Visual Accelerometer.

Step 4: Which statement best describes the connection between the LEDs and the motion of the dynamics cart that it's attached to?

____The LEDs indicate the <u>position</u> of the cart: the green ones light whenever the cart is on the right side of the track, and the red ones light whenever the cart is on the left. The greater the distance to either side, the greater the number of LEDs that light.

____The LEDs indicate the <u>velocity</u> of the cart: the green ones light whenever the cart moves right, and the red ones light whenever the cart moves left. The greater the speed, the greater the number of LEDs that light.

____The LEDs indicate the <u>acceleration</u> of the cart: the green ones light whenever the cart accelerates to the right, and the red ones light whenever the cart moves to the left. The greater the acceleration, the greater the number of LEDs that light.

PART B: WHICH WAY IS THE "A"?

Step 1: Set one Visual Accelerometer near the edge of a rotational platform, along a *tangent* (not a radius). Make sure it's turned on and the range is set for "Manual: 5 m/s²."

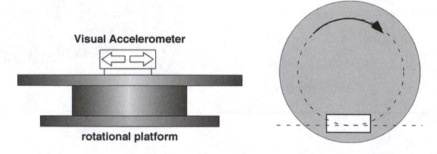

Step 2: Prediction. What will happen when the platform is rotated *clockwise* and the Visual Accelerometer goes into circular motion?

__the green (rightward) LEDs will light

__the red (leftward) LEDs will light

__all LEDs will light

__no LEDs will light

Step 3: Observation. What actually happens when the platform is rotated, and what does it mean for the *tangential* acceleration of an object in uniform circular motion? While in clockwise uniform circular motion, is there acceleration forward (clockwise), backward (counterclockwise), or essentially no tangential acceleration at all?

_____

_____

Step 4: Reorient the Visual Accelerometer so that it lies along a *radius* of the platform (instead of on a *tangent* as before).

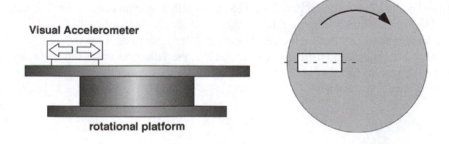

Step 5: Prediction. What will happen when the platform is rotated clockwise and the Visual Accelerometer goes into circular motion?

___the green (rightward) LEDs will light

___the red (leftward) LEDs will light

___all LEDs will light

___no LEDs will light

Step 6: Observation. What actually happens when the platform is rotated, and what does it mean for the *radial* acceleration of an object in uniform circular motion? While in clockwise uniform circular motion, is there acceleration inward (toward the center), outward (away from the center), or essentially no radial acceleration at all?

_____

_____

Step 7: Prediction. What will happen when the platform is rotated *counterclockwise* and the Visual Accelerometer goes into circular motion?

_____

Step 8: Observation. What actually happens when the platform is rotated, and what does it mean?

_____

_____

Step 9: Reposition the Visual Accelerometer so that its center is at the center of the rotational platform.

Step 10: Prediction. What will happen when the platform is rotated clockwise and the Visual Accelerometer goes into rotational motion?

_____

Step 11: Observation. What actually happens when the platform is rotated, and what does it mean?

_____

_____

Will it Go 'Round in Circles?

Step 12: Arrange two Visual Accelerometers along a diameter of the rotational platform. Place each one at the midpoint of the radius it's on.

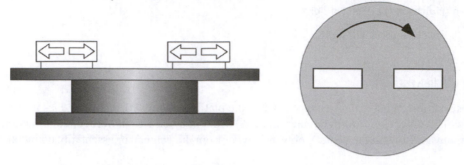

Step 13: Prediction. What will happen when the platform is rotated clockwise and the Visual Accelerometers go into rotational motion?

_____

Step 14: Observation. What actually happens when the platform is rotated, and what does it mean?

_____

_____

Step 15: Conclusion. Which conclusion best matches your observations?

The acceleration of an object in uniform circular motion

____ is tangentially forward (in the direction of motion along the tangent).

____ is tangentially backward (opposite the direction of motion along the tangent).

____ is radially inward (toward the center of the circle).

____ is radially outward (away from the center of the circle).

____ does not exist.

PART C: THE CAROUSEL CONUNDRUM

Step 1: Arrange two Visual Accelerometers along a diameter of the rotational platform. Place one nearer to the center and the other one farther from the center.

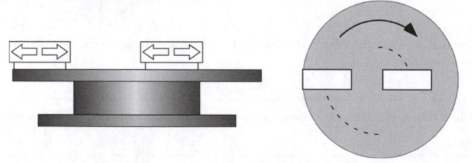

Step 2: Prediction. What will happen when the platform is rotated? Specifically, which accelerometer, if either, will undergo *greater* acceleration?

_____

Step 3: Observation. What actually happens when the platform is rotated, and what does it mean?

_____

_____

Step 4: To better understand the observation, answer the following questions.

a. Which accelerometer, if either, travels a greater *distance* during each revolution of the platform?

_____

b. Which accelerometer, if either, requires more *time* to complete a revolution?

_____

c. Which accelerometer, if either, travels with greater linear *speed* while the platform rotates?

_____

d. The equation for the acceleration on an object in uniform circular motion is $a = v^2/r$, where v represents linear speed and r represents orbital radius (the distance from the center of the circle).

Summing Up
PART B: WHICH WAY IS THE "A"?
1. What is the difference between the terms "centripetal" and "centrifugal"?

_____

_____

_____

2. Which term is more appropriate to describe the acceleration you observed when the accelerometer was in uniform circular motion?

_____

PART C: THE CAROUSEL CONUNDRUM
3. Explain the observation in Part C, Step 3 in terms of your findings in Part C, Step 4.

_____

_____

_____

Going Further
PART D: NEWTON'S CRADLE
Step 1: Remove the Visual Accelerometers and set them aside. Carefully place a Newton's cradle on the platform so that it is centered at the platform's center. The center ball should lie on the platform's axis of rotation.

Newton's cradle

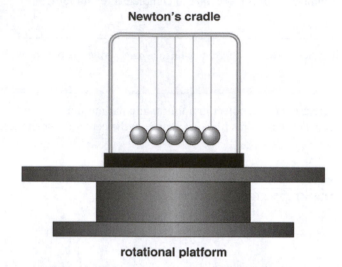

rotational platform

Step 2: Prediction. What, if anything, will happen to the cradle when the platform is rotated? Predict using words and a diagram.

_____

_____

Step 3: Observation. What actually happens to the cradle when the platform is rotated? Describe using words and a diagram.

_____

_____

Step 4: To better understand this observation, consider the following points.

a. Consider one ball suspended by a cord some distance from the axis of rotation. If the platform is rotated, the ball will undergo uniform circular motion.

b. A ball in uniform circular motion requires something to push or pull it toward the center of the circle. Tension in the cord can provide this force. But only if the cord is at an angle.

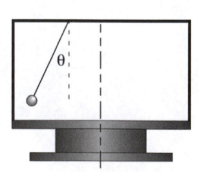

c. Draw the forces that act on the suspended ball when it's in uniform circular motion. One is gravitational force (mg), the other is tension (T).

d. The vertical forces balance: $T\cos\theta = mg$

e. The horizontal force is the centripetal force: $T\sin\theta = mv^2/r$

f. In the space below, solve for θ using only m, g, v, r, and constants. (You may not need all the variables listed.

Step 5: Use your result from Step 4 to answer the questions that follow.

How—if at all—would the angle (θ) have been different if

a. the ball had more mass?

_____

b. the experiment had been conducted on the Moon?

_____

c. the platform were spinning faster?

_____

d. the radial distance were greater?

_____

Step 6: Remove the Newton's cradle from the platform.

PART E: WATER WHIRLED

Step 1: Fill the Laser Viewing Tank halfway with water.

Step 2: Add several drops of food coloring to make the water visible.

Step 3: Place the tank (or liquid accelerometer) on the rotational platform so that its center is on the platform's axis of rotation. Secure the tank to the platform (with duct tape, perhaps).

Step 4: Prediction. What will happen to the water when the platform is rotated? Predict using words and a diagram.

_____

_____

Step 5: Observation. What actually happens when the platform is rotated? Describe using words and a diagram.

_____

_____

Will it Go 'Round in Circles?

CONCEPTUAL PHYSICS	**Activity**

Sit on It and Rotate

Purpose
To experience conservation of angular momentum by feeling its effects

Apparatus
low-friction rotating stool
2 dumbbells (or equivalent, such as bricks)
1 bicycle wheel with handles
an assistant (instructor or lab partner, for example)

SAFETY CAUTION: BE CAREFUL AROUND THE SPINNING BICYCLE WHEEL. KEEP FINGERS AND DANGLING OBJECTS (LONG HAIR, JEWELRY, LOOSE CLOTHING, ETC.) AWAY FROM THE SPOKES.

Discussion
Three quantities in the universe are the same now as they ever have been or ever will be. One is energy. Another is linear momentum. The third is angular momentum. In the absence of external torques, the angular momentum of a system is conserved. In this activity, you will feel what that statement means. You will also learn some implications of that principle.

Procedure
PART A: THE ICE CAPS MELTETH
1. If the polar ice caps were to melt, a considerable mass of water would flow from the poles toward the equator. Would this tend to make days longer or shorter? Explain.

Step 1: Sit on the rotating stool so that you are balanced and comfortable. Obtain the dumbbells; hold one in each hand. Raise the dumbbells over your head and lift your feet off the ground.

Step 2: Let your assistant apply a gentle torque to get you rotating no faster than one revolution per second.

Step 3: Let the ice caps melt: While keeping your arms fully outstretched, move the dumbbells from over your head to straight out (see Figure 1). Bring the dumbbells back up and repeat so you have a clear sense of the changes you feel, if any, as your arms move.

Figure 1.a. Ice Cap Frozen; Dumbbells over Head

Figure 1.b. Ice Cap Melting

Figure 1.c. Water Fully Redistributed

Figure 1.d. Ice Cap Freezing Back Up for Repeat

a.   What happens to your angular speed (rate of rotation) as you move the dumbbells from the "poles" to the "equator"?

_____

b.   How would the length of the day be affected if the polar ice caps were to melt?

_____

PART B: SLICK SPOT

Suppose you're riding a bicycle and you suddenly find yourself on a slick spot, such as a patch of ice or an oil slick on the roadway. Should you keep pedaling or lock the wheels as you slide across the slick spot? To find out, complete the following activity.

Step 1: Hold the nonrotating bicycle wheel by its handles. Orient the bicycle wheel vertically (as it would be if you were riding the bicycle).

Step 2: Quickly twist the wheel into a horizontal orientation (as it would be if you were to fall). See Figure 2.

Figure 2a. The Bicycle Wheel, Vertical

Figure 2b. The Bicycle Wheel, Horizontal

Figure 2c. Horizontal with the other Handle Up

Step 3: Rotate the bicycle all the way over (180°) so that it is horizontal with the other handle up.

Step 4: Reorient the wheel vertically. Have your assistant give the wheel a significant torque so that the wheel spins rapidly in the vertical orientation.

Step 5: Now repeat steps 2 and 3: twist the rotating wheel to one horizontal orientation and then back over to horizontal with the other handle up.

Step 6: It's safe to slow the wheel by gently pressing the palm of your hand to the rubber tread of the wheel. Stopping the wheel on the floor often leaves a mark.

Step 7: Allow your assistant to slowly bring the wheel to a stop by applying *gentle* pressure to the outer surface of the tire with the palm of his or her hand.

a. When is it harder to "wipe out" (fall): when the wheel is at rest or when the wheel is spinning?

_____

_____

b. What's the best strategy for remaining upright when you hit a slick spot on the road?

_____

_____

PART C: THE GO SPIN ZONE
Conservation of angular momentum has curious consequences when you use the bicycle wheel and the rotational stool together.

Step 1: Sit on the rotational stool so that you are balanced and comfortable.

Step 2: Have your assistant give you the bicycle wheel. Hold it vertically. See Figure 3.

Step 3: Keep your body stationary (keep your feet on the ground, if possible), keep your arms rigid, and hold the bicycle wheel tightly while your assistant applies a significant torque to the wheel to get it spinning.

Note: If viewed from overhead, there would be no angular momentum in the system (you on the rotational stool holding the bicycle wheel). No spin could yet be observed. See Figure 4.

Step 4: Without pushing off, carefully free your body by lifting your feet. You should not be rotating at this point.

Step 5: Turn the spinning bicycle wheel from a vertical orientation to a horizontal orientation.

What happens when you turn the bicycle wheel on its side?

_____

_____

Figure 3

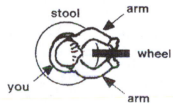

Figure 4. As Seen from Overhead

Step 6: Turn the bicycle wheel all the way over so that it is horizontal with the other handle up.

a. If seen from overhead, are you spinning clockwise or counterclockwise? Answer below and add an arrow showing your spin to Figure 5.

_____

b. If seen from overhead, is the bicycle wheel spinning clockwise or counterclockwise? Answer below and add an arrow showing the wheel's spin to Figure 5.

![Figure 5 overhead view of person and wheel]

Figure 5. As Seen from Overhead.

Sit on It and Rotate

c. Which object spins with greater angular *speed:* you or the bicycle wheel?

_____

d. Which object has more rotational *inertia:* you or the bicycle wheel?

_____

e. If seen from overhead, which object has more *angular momentum:* you or the bicycle wheel?

_____

f. Before the bicycle wheel was turned on its side, there was a total of *zero* angular momentum in the system. What is the total angular momentum in the system when the bicycle wheel is turned on its side? Explain.

_____

_____

Summing Up

1. The Mississippi River flows from north to south along the surface of the Earth. It picks up solid material and deposits it in the delta, where the river meets the Gulf of Mexico. What is the effect of this deposition on the rate of rotation of the Earth? And what effect does this have on the length of the day?

_____

_____

2. River reservoirs hold back substantial masses of water that would otherwise flow toward the equator. What is the effect of this retention on the rate of rotation of the Earth? And what effect does this have on the length of the day?

_____

_____

3. Suppose you are spinning on the rotating stool with arms outstretched and dumbbells in your hands. And suppose you pull the dumbbells in toward your chest.

   a. What happens to your angular speed as you pull the dumbbells in?

   _____

   b. What happens to your rotational inertia as you pull the dumbbells in?

   _____

   c. What happens to your angular momentum as you pull the dumbbells in?

   _____

Name _____ Section _____ Date _____

| CONCEPTUAL PHYSICS | Demonstration |

The Big BB Race

Purpose
To compare the path of a projectile launched horizontally with that of an object in free fall

Apparatus
simultaneous launcher/dropper
table clamp   support rod   right-angle clamp

Discussion
Suppose a ball bearing (BB) were launched horizontally at the same time another BB were dropped from the same height. Which one would reach the ground first?

Procedure
Step 1: Draw the path you think the launched BB will take on Figure 1. That is, draw a line connecting the launch point and the impact point in the diagram that traces the path you think the BB will follow.

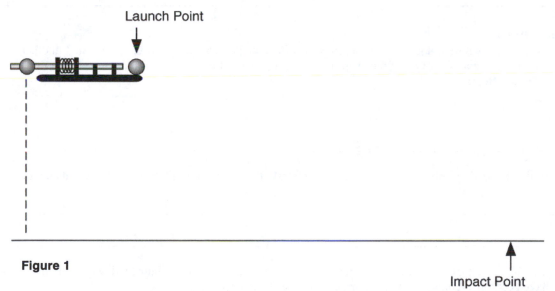

Figure 1

Step 2: Consider the following predictions of three students. Write arguments supporting each prediction (whether you agree with the prediction or not).

a.   Student X predicts the dropped BB will hit first. Why might Student X believe this?

_____

b.   Student Y predicts the fired BB will hit first. Why might Student Y believe this?

_____

c. Student Z predicts the dropped BB will hit first. Why might Student Z believe this?

Step 3: One of the reasons sometimes offered to support the prediction that the dropped ball will hit first is that the launched ball will travel forward for some distance before starting to move downward. What factors might determine the length of this "no-fall distance," as shown in Figure 2?

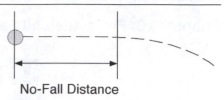

No-Fall Distance

Figure 2

Step 4: Which prediction do you agree with: dropped BB hits first, launched BB hits first, or both hit at the same time?

Step 5: Perform and observe the operation of the simultaneous launch/drop mechanism.

a. Observe a dropped BB.

b. Observe a launched BB.

c. Observe The Big BB Race—a simultaneous launch **and** drop.

Step 6: Which BB hits the ground first, or is it a tie?

Summing Up

1. How does the horizontal motion of a projectile affect the vertical motion of the projectile? In other words, does the horizontal motion of the projectile make it move faster or slower in the vertical direction (or does it have no effect)?

2. Which factors—if any—appear to have the greatest effect on the no-fall distance discussed above?

3. If the launched BB had a rocket engine propelling it forward after it was launched, what—if anything—would have been different about the outcome of The Big BB Race?

CONCEPTUAL PHYSICS

Experiment

Bull's Eye

Purpose
To predict the landing point of a projectile

Apparatus
1/2" (or larger) steel ball   table
empty can   meterstick
stopwatch
means of projecting the steel ball horizontally at a known velocity

Discussion
Figures 10.5–10.8 in *Conceptual Physics* show how projectiles move with constant speed in the horizontal direction while undergoing free-fall acceleration in the vertical direction. Figure 1 at the right shows the same for a ball tossed horizontally. The horizontal and vertical motions are ***independent*** of each other!

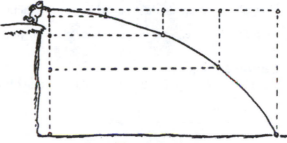

Figure 1

Instead of tossing a ball from a cliff, you'll fire a steel ball off the edge of a table and into a can on the floor below—all without a trial shot!

When engineers build bridges or skyscrapers, they do ***not*** do so by trial and error. For the sake of safety and economy, the effort must be right the ***first*** time. Your goal in this experiment is to predict where a steel ball will land when projected horizontally from the edge of a table. The final test of your measurements and computations will be to position an empty can so that the ball lands in the can on the ***first*** attempt.

Procedure
Step 1: Your instructor will provide you a means of projecting the steel ball at a known horizontal velocity (it may be a spring gun, a ramp, or some other device). Position the ball projector at the edge of a table so the ball will land downrange on the floor. Do ***not*** make any practice shots (the fun of this experiment is to ***predict*** where the ball will land without seeing a trial). Figure 2 shows the quantities involved in this experiment.

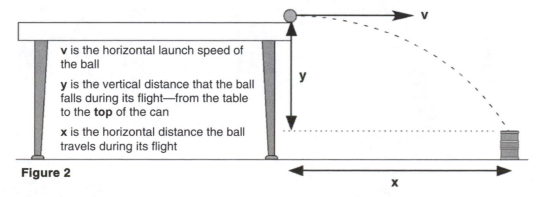

v is the horizontal launch speed of the ball

y is the vertical distance that the ball falls during its flight—from the table to the **top** of the can

x is the horizontal distance the ball travels during its flight

Figure 2

You will be given the initial speed of the steel ball (or perhaps you'll have to devise a way to measure it). Record the firing speed here.

Horizontal speed v = _____ cm/s

Step 2: Carefully measure the vertical distance y the ball must drop from the bottom end of the ball projector in order to land in an empty soup can on the floor. Be sure to take the height of the can into account when you make this measurement.

Vertical distance y = _____ cm

Step 3: The vertical free-fall motion is governed by the equation $y = (1/2)gt^2$. A rearrangement of the equation, $t = \sqrt{(2y/g)}$, lets you calculate the time it takes the ball to fall from its original height. For gravitational acceleration, use $g = 980$ cm/s^2.

Time of flight t = _____ s

Step 4: The range is the horizontal distance a projectile travels, x. Predict the range of the ball using $x = vt$. Write down your predicted range.

Predicted range x = _____ cm

Now place the can on the floor where you predict it will catch the ball.

Step 5: Only after your instructor has checked your predicted range and your can placement, shoot the ball.

Summing Up

1. Did the ball land in the can on the first trial? If not, how many trials were required?

_____

2. What possible errors would account for the ball overshooting the target?

_____

3. What possible errors would account for the ball undershooting the target?

_____

4. If the can you used were replaced with a taller can, would you get a successful run if the ball started with a higher speed, lower speed, or the same speed? Explain.

_____

_____

| CONCEPTUAL PHYSICS | Experiment & Tech Lab |

Blast Off!

Purpose

To measure the range of an air-powered projectile and then replicate the motion using a projectile motion simulation to determine the launch speed of the projectile

Apparatus

Arbor Air-Powered Rocket body
Arbor Thrust Washers
Arbor Nose Cone
Arbor Launch Pad
Arbor Angle Wood Wedges
air pump
meterstick
access to electronic balance
football field or
    300-ft measuring tape or
    trundle wheel

computer
PhET simulation: "Projectile Motion"
(at http://phet.colorado.edu)

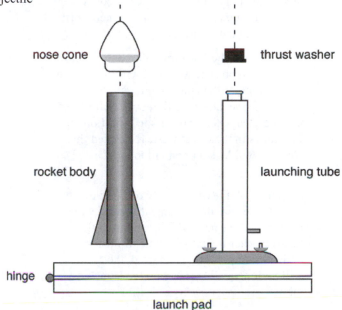

SAFETY NOTE: THE AIR-POWERED PROJECTILES ARE FOR OUTDOOR USE ONLY. USE EYE PROTECTION WHEN LAUNCHING THE AIR-POWERED PROJECTILES.

ROCKETS LAUNCH WITH HIGH SPEED! KEEP CLEAR OF THE ROCKET'S PATH. DO NOT LAUNCH TOWARD PEOPLE OR ANIMALS.

Discussion

The physics of projectile motion eluded great minds for centuries. Galileo observed that the path of a projectile could be described as a parabola. Galileo's solution to the parabolic projectile puzzle was an important step on his path to understanding inertia.

In this experiment and tech lab, you will first launch an air rocket at three different angles and record the corresponding ranges. You will then model your result using a computer simulation based on Galileo's solution to the projectile puzzle. By doing so, you will be able to determine the launch speed of your air rocket. Lastly, you'll use the simulation to see how far your rocket would have traveled in the absence of air resistance.

Procedure

PART A: FIELD WORK—WORKING WITH THE AIR ROCKETS

The Air-Powered Projectile uses air pressure to provide thrust to a lightweight rocket. Different "thrust washers" can be used to vary the pressure at which the rocket launches, thereby varying the launch speed. Wood wedges can be used to vary the launch angle.

Step 1: Carefully measure the diameter and mass of the rocket. The diameter is measured at the widest point of the projectile: the nose cone. Record the diameter in *meters* and the mass in *kilograms*.

d = _____   m = _____

Step 2: Gather the following materials and bring them to the launch site: launch pad with launching tube, air pump, thrust washers, rocket body, and nose cone. If the launch site is not a football field, bring a measuring tape or trundle wheel.

Step 3: Connect the air pump to the launching tube at the brass fitting. Secure the connection. (Some air pumps have a clamping mechanism that secures the connection.)

press to fit

Step 4: Press the *low* thrust washer onto the launching tube. The wide end will fit over the top opening of the launching tube. This is a friction fit; securing the fit requires effort and the knowledge that the parts *will* fit. Look out for cracks in the thrust washer; cracked thrust washers will leak air and fail to launch properly.

press to fit

Step 5: Push the rocket body onto the thrust washer. Be sure the alignment is good while pressing. The small cylinder on top of the thrust washer fits into the hole of the rocket body. When properly connected, the rocket will no longer slide freely on the launching tube.

"massage" to fit
must be airtight

Step 6: Now attach the nose cone. This task requires effort, skill, and patience. The rubber cone fits in and around the top of the rocket body. When it's seated correctly, a *small* squeeze of the cone is met with resistance: The seal is airtight and will bulge where not being squeezed. So if squeezing the cone deflates it, the seal is not set. The cone must be seated correctly for a smooth flight and a safe landing.

Step 7: Set the 35° launch angle wedge in the launch pad slot. The geometry is tricky here, so pay attention to the details: The 35° *launch angle* wedge has one corner cut at *55°*. This assures a launch that is 35° above horizontal. Close the hinged launch pad with the wedge as far forward as possible. There must be no gap between the angled side of the wedge and the launch pad.

angle wedge fits in slot (move forward)

Step 8: Place the air pump behind or to the side of the launch pad. The pump hose is not very long, but you want as much distance as possible from the rocket for a safe launch.

Step 9: Clear the immediate area and alert bystanders to the imminent launch with an audible warning (e.g., calling out "Fire in the hole!"). Do not proceed until you have a clear corridor in front of you and your instructor's permission.

Step 10: The air pump operator *must* wear eye protection. From a position behind and away from the launch pad, pump air into the launching tube with repeated strokes of the air pump. Keep pumping until the rocket launches.

The rocket will launch when the pressure reaches a specific value.

Step 11: Note the impact point and measure range: the distance from the launch pad to the point of impact. Record the value on the data table. Note: The first column for range has been left without the units specified. If you are using a football field, record the values in yards. If you are using a measuring tape labeled in feet, use feet. Be sure to record the units you're using in the column heading. If your measurements of range are made in meters, leave that column blank and record your values directly in the second range column. The blank columns will be used in Part B.

Step 12: Prepare for the next launch.
a. Remove the thrust washer from the retrieved rocket body.
b. Repeat Steps 4–6.
c. Replace the 35° launch angle wedge with the 45° wedge.

Step 13: Launch. Repeat Steps 8–12.

Step 14: Repeat for a launch angle of 55°.

Step 15: Carry out one final launch. Replace the *low* thrust washer with the *super* thrust washer. Select the angle that, based on your findings, will result in the greatest range. Record the launch angle and range in the bottom row of the data table.

Step 16: Convert the range values to meters and record the converted values in the column titled "Range R (m)." See Appendix D for conversion factors.

Data Table

Angle θ (°)	Range R (___)	Range R (m)			
35					
45					
55					

PART B: COMPUTER WORK—MODELING ROCKETS

SETUP

Step 1: Turn on the computer and allow it to complete its start-up cycle.

Step 2: Open the PhET simulation, "Projectile Motion." If you need assistance, ask your instructor for help.

Step 3: Size the screen; size the scene.

a. Stretch the window so that it fills the screen.

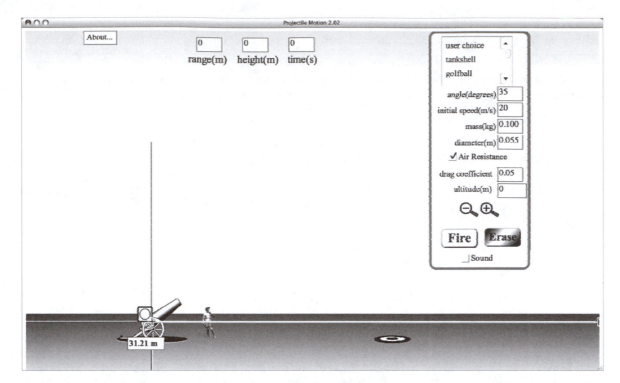

PhET's "Projectile Motion"

b. Drag the on-screen tape measure to the intersection of the "crosshairs" that intersect near the top of the cannon wheel. Pull the free end of the tape horizontally to the right until it reaches the edge of the window. Note the length displayed below the tape measure base.

c. Find the greatest range value recorded on the data table. Click the on-screen "Zoom Out" button (a magnifying glass with a *minus* sign) until the length of the tape is greater than the greatest range value you recorded.

Step 4: From the object list at the top of the on-screen control panel, choose "tankshell."

Step 5: In the angle field of the on-screen control panel, enter 35. In the speed field, enter 20. In the mass field, enter the mass of the air rocket as recorded in Part A, Step 1. In the diameter field, enter the diameter of the projectile as recorded in Part A, Step 1.

Step 6: Click the checkbox in the on-screen control panel to activate "Air Resistance."

Step 7: In the altitude field, enter the elevation above sea level of your location. Ask your instructor for assistance determining this value.

Elevation above sea level (altitude) = _____ m

MATCHING THE RESULTS
Step 8: Fire the cannon! Use the tape measure to determine the horizontal range. Keep the fixed end of the tape measure at the crosshairs and drag the free end to the plotted path of the projectile.

Step 9: Compare the range of the simulated projectile with that of the air rocket you fired with a 35° launch angle.

Step 10: Modify the launch speed and fire again until your simulated projectile's range matches your air rocket's range (as nearly as possible).

Step 11: In the data table, label the first empty column "Sim Speed v (m/s)."

Step 12: Once you find a match at 35°, change the launch angle to 45°. Through trial and error, find the correct launch speed so that the simulation range matches the measured air rocket range.

Step 13: Find the matching launch speed at 55°. Repeat for the "maximum range" final launch of the air rocket.

IF IN A VACUUM
Air resistance robs a projectile of height and range. With the simulation, you can see how far the projectile would have traveled in the absence of air resistance.

Step 14: In the data table, label the next empty column "Vacuum Range R* (m)."

Step 15: Click the on-screen checkbox to deactivate "Air Resistance."

Step 16: Re-create each of your launches, starting with the 35° launch. Note and record the corresponding range.

Step 17: In the data table, label the final blank column "% of Max. Range."

Step 18: Divide each actual air rocket range by the corresponding vacuum range and multiply the result by 100. For example, an actual range of 30 m and a vacuum range of 37 m would produce a result of $\frac{30 \text{ m}}{37 \text{ m}} \times 100 = 81\%$.

Summing Up

1. In a vacuum, a projectile launched at 35° would have the same range as one launched at 55°. Which actual rocket traveled farther when you launched them in air?

    _____

    _____

2. You made two launches at the same angle: one with the *low* thrust washer, the other with the *super* thrust washer.

    a. Which—if either—was greater as a result: the increase in *speed* or the increase in *range?* Answer by calculating the ratio of super to low launch speeds and super to low ranges.

    _____

    _____

    b. Which range was affected more by air resistance?

    _____

3. In the simulation, the tankshell has a very low drag coefficient. That is, it is very aerodynamic and is affected only slightly by air resistance.

    a. Which object in the simulation has the *greatest* drag coefficient?

    _____

    b. If launched at the speed and angle simulating your maximum range air-rocket launch, how far would that high-drag object go? And what percent of maximum range would that be?

    _____

    _____

Blast Off!

4. Based on your findings, which type of launch would suffer *most*—in terms of range—from air resistance?

____ low and slow (small launch angle and low launch speed)

____ low and fast

____ high and slow

____ high and fast

5. What would be the effect on projectile range if these launch angles and speeds were repeated on the Moon?

_____

_____

6. (Optional). How is David's modesty maintained even if his boxers are removed?

_____

CONCEPTUAL PHYSICS	Tech Lab

Worlds of Wonder

Purpose
To use a simulation to study the orbital mechanics of a simplified solar system

Apparatus
computer
PhET simulation: "My Solar System" (available at http://phet.colorado.edu)

Discussion
Lab activities involving stars and planets are difficult to conduct inside a classroom or laboratory. Because you cannot create stars and planets to experiment with in the classroom, you will use a computer simulation that uses the laws of gravity to show the behavior of large objects at great distances from one another.

Procedure

SETUP
Step 1: Turn on the computer and let it complete its start-up process.

Step 2: Open the PhET simulation, "My Solar System." If you're not sure how to do this, ask your instructor for assistance.

Step 3: When the simulation opens, the screen should resemble the figure below.

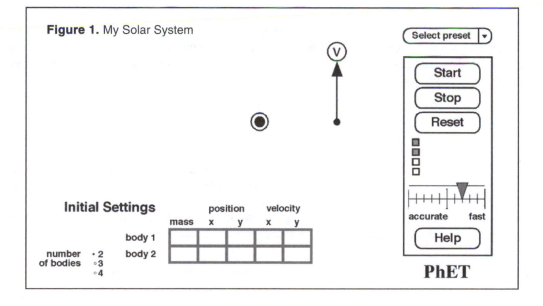

PART A: NEWTON'S CANNON

Isaac Newton explained that universal gravitation accounted for both the fall of an apple and the orbit of the Moon. At the time, this was hard for people to understand. Newton used a thought experiment to show how the same force could explain free-fall and orbital motion. In this activity, you will simulate "Newton's Cannon."

Step 1: In the control panel on the right side of the screen, the checkboxes for "System Centered" and "Show Tracks" should be checked. Set the "accurate/fast" slider to the midpoint. Set the "Initial Settings" for body 1 (yellow Sun) and body 2 (pink planet) as follows.

      a. Body 1: mass = 200, position x = 0, position y = 0, velocity x = 0, velocity y = 0.

      b. Body 2: mass = 1, position x = 0, position y = 100, velocity x = 0, velocity y = 0.

Step 2: Click the on-screen "Start" button and record your observation.

_____

_____

Step 3: Click the on-screen "Reset" button to stop the simulation and restore the initial position and velocity settings.

Step 4: Change the initial Velocity x of body 2 (the pink planet) to 40. Click the on-screen "Start" button and record your observation of what happens. How is it different from your previous observation?

_____

_____

Step 5: Click the on-screen "Reset" button to stop the simulation and restore the initial position and velocity settings. Change the initial Velocity x of the pink planet to 80. Click the on-screen "Start" button and record your observation.

_____

_____

Step 6: Click the on-screen "Reset" button. Change the initial Velocity x of the pink planet to 160. Click the on-screen "Start" button and record your observation of what happens. How is it different from your previous observation?

_____

_____

Step 7: Click the on-screen "Reset" button. Through trial and error, determine the minimum initial Velocity x that will allow the pink planet to orbit the yellow Sun. From your previous investigations, you know a speed of 40 is too small and an initial speed of 80 is more than enough. So your result will be between 40 and 80. Don't worry if the animation shows the planet moving through the Sun. What is the minimum initial Velocity x that will allow the pink planet to orbit the yellow Sun at least 10 times without crashing?

_____

Step 8: Click the on-screen "Reset" button. On the control panel, click the "Show Grid" checkbox. Through trial and error, determine the correct initial Velocity x that will allow the pink planet to orbit the yellow Sun in a *circular* orbit. If the initial speed is too high or too low, the orbit will be elliptical. What speed is just right to allow a *circular* orbit?

PART B: HARMONY OF THE WORLDS

There is a mathematical relationship between the orbital radius and orbital speed of planets circling the Sun. German mathematician Johannes Kepler discovered this relationship. He started with volumes of astronomical data, worked through hundreds of pages of calculations, and spent approximately 30 years pursuing the discovery. In this activity, you'll use the simulation to generate data that will allow you to make the discovery in much less time.

Step 1: Find circular orbits for planets at various distances from the Sun. Start by setting the "position y" of the pink planet at a distance of 50. This sets the orbital radius to 50.

Step 2: On the control panel, click to activate the "Tape Measure."

Step 3: Click and drag the tape measure box icon until its crosshairs (+) are on the pink planet. Now click and drag the other end of the tape measure vertically downward, across the Sun, until it measures a distance of 100. Because you set the orbital radius to 50, the orbital diameter is 100. So the tape measure represents the diameter of the orbit.

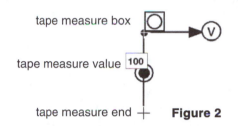

Figure 2

Step 4: Set the Velocity x of the pink planet to 150. Click the on-screen "Start" button and observe the orbit. Because the trace of the pink planet doesn't pass through the far end of the tape measure, the orbit is not circular.

Step 5: Click the on-screen "Reset" button. Try a different Velocity x for the pink planet. Through trial and error, keep trying until you find the speed that results in a circular orbit. The trace of the pink planet will pass through the far end of the tape measure when the orbit is circular. Record the Velocity x on the data table.

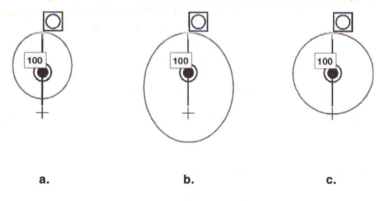

a.   b.   c.

Figure 3
a. and b. Noncircular Elliptical Orbits c. Circular Orbit

Worlds of Wonder

Step 6: Find circular orbits when the orbital radius is 100, 150, and 200 to complete the data table.

Data Table

Orbital Radius R (Position y)	Orbital Speed v (Velocity x)
50	
100	
150	
200	

Summing Up

PART A: NEWTON'S CANNON

1. A cannonball dropped from a cliff will fall straight down and hit the surface of the Earth. How could the cannonball be made to orbit the Earth instead?

_____

_____

2. Based on your experience with the simulation, which do you think is more common: *circular* orbits or noncircular *elliptical* orbits? Defend your answer.

_____

_____

PART B: HARMONY OF THE WORLDS

3. Use the following method to determine the relationship between the orbital radius of a planet and the orbital speed of its circular orbit. For this activity, you'll limit your investigation to three possible relationships. They are as follows:

Orbital radius is inversely proportional to orbital speed: $R \sim 1/v$.

Orbital radius is inversely proportional to the square of orbital speed: $R \sim 1/v^2$.

Orbital radius is inversely proportional to the square root of orbital speed: $R \sim 1/\sqrt{v}$.

a. To see the pattern in the data, simplify and process your data. First, rewrite the orbital data on the table below.

R	v	R*	v*	1/v*	1/v*²	1/√v*
50		1.00	1.00	1.00	1.00	1.00
100		2.00				
150						
200						

b. Divide each value in the Orbital Radius column by the first value in the Orbital Radius column (50). Record the results in the R* column of the table above. That is, the values in the R* column will be the results of the quotients 50/50, 100/50, 150/50, and 200/50.

c. Repeat this process using the Orbital Speed data to determine values of v*. That is, divide all values of Orbital Speed by the first value of orbital speed.

d. Now complete the last three columns by performing the appropriate mathematical operations on the values in the v* column.

4. Select the column that best matches the R* column. Is it ____1/v*, ____1/v*², or ____1/√v*?

5. Complete the statement: **Orbital radius is inversely proportional to the**

_____

> Johannes Kepler worked out the mathematics of orbits. Isaac Newton used Kepler's findings to develop the Theory of Universal Gravitation!

CONCEPTUAL PHYSICS　　　　　　　**Tech Lab**

Part One: Mechanics　　　　　　　　　　Qualitative Video Analysis

Bicycle Dancer of Edinburgh

Purpose

To analyze a short video of a street trials bicycle performer, identifying the physics involved in the stunts he performs.

Apparatus

computer with Internet access
YouTube video v=Z19zFlPah-o ("Inspired Bicycles – Danny MacAskill April 2009")
headphones and signal splitters (optional)

Discussion

The web video clip, "Inspired Bicycles – Danny MacAskill April 2009," is a professionally produced film featuring the bike stunts of a talented world-class performer. The stunts are feats of skill, talent, and outstanding demonstrations of physics!

The description of the 5 1/2-minute video is as follows: *Filmed over the period of a few months in and around Edinburgh by Dave Sowerby, this video of Inspired Bicycles team rider Danny MacAskill (more info at www.dannymacaskill.com) features probably the best collection of street/street trials riding ever seen. There's some huge riding, but also some of the most technically difficult and imaginative lines you will ever see. Without a doubt, this video pushes the envelope of what is perceived as possible on a trials bike. Credit to Band of Horses for their epic song "The Funeral."*

Procedure

Step 1: Turn on the computer and allow it to complete its start-up cycle. If you are completing this activity in a laboratory or classroom setting, connect signal splitters and headphones to the computer as needed.

Step 2: Launch the computer's web browser (e.g., Firefox, Safari, Internet Explorer).

Step 3: Locate the YouTube video, "Inspired Bicycles – Danny MacAskill April 2009." It should be located at http://www.youtube.com/watch?v=Z19zFlPah-o. That is, it is video Z19zFlPah-o within YouTube's vast collection.

Step 4: Watch the video in its entirety. Mute the sound unless you are using headphones or completing this activity in a private setting. It's preferable to watch the first time with the sound on.

Step 5: When the video has played to its conclusion, set the computer aside and complete the steps that follow.

Step 6: Review the following vocabulary by matching each term to its definition.

Gravitational Potential Energy •　　• energy of motion

Chemical Potential Energy •　　• interval during which a collision takes place

Kinetic Energy •　　• delicate balance

Centripetal •　　• energy of position (based on height above ground)

Unstable Equilibrium •　　• directed toward the center of a circle or arc

Impact Time •　　• energy of position (based on molecular interaction)

Step 7: Match each energy transformation process to the correct description.

Turning chemical potential energy into · gravitational potential energy   · trading speed for height

Turning chemical potential energy into · kinetic energy   · using muscles and food energy to gain height

Turning kinetic energy into gravitational · potential energy   · trading height for speed

Turning gravitational potential energy into · kinetic energy   · using muscles and food energy to gain speed

Step 8: Match each term to the corresponding description.

Dynamic Unstable Equilibrium ·   · spinning as in a somersault

Centripetal Acceleration ·   · maintaining delicate balance while moving

Rotation Around a Vertical Axis ·   · "cushions" a collision by "absorbing" the effects

Rotation Around a Horizontal Axis ·   · spinning as a helicopter blade would spin

Impact: increase time to decrease force ·   · the change in velocity required for circular motion

Step 9: Examine the checklist on the following page. Each stunt shown in the video clip involves one or more principles of physics from the checklist. Each stunt has been given a title on the checklist, such as "Death Fence" or "Bike Shop—Copy Stop Hop."

Step 10: Watch the video clip again. Pause and rewind the playback as needed. ***Identify the primary principles demonstrated in each stunt.*** A single stunt may involve up to five principles. Many involve at least three; some involve only one. Place a checkmark in the appropriate column or columns for each stunt. ***When you are done, each of the 26 rows will have at least one checkmark and each of the 9 columns will have at least one checkmark.***

SAFETY NOTE: DANNY MACASKILL IS A PROFESSIONAL RIDER WITH OVER 12 YEARS OF TRAINING. MANY STUNTS PERFORMED IN THE VIDEO CLIP ARE EXTREMELY DANGEROUS. DO NOT ATTEMPT ANY OF THESE STUNTS ON YOUR OWN!

Summing Up

1. During which stunts does MacAskill ride *without* his helmet?

2. Estimate the maximum potential energy that MacAskill attains when riding up the tree in Stunt 2.

    a. Which two quantities must you estimate to carry out the calculation?

    b. Estimate those two quantities and record your estimates.

    c. Carry out the calculation and record your result.

	Chemical PE to Grav PE	Chemical PE to Kinetic E	Kinetic E to Grav PE	Grav PE to Kinetic E	Dyn Unstable Equilibrium	Centripetal Acceleration	Rotation Vert Axis	Rotation Horiz Axis	Increase ↑ Decrease ↓ F
1. "Death" Fence									
2. Up a Tree									
3. Barricade Hop (x2)									
4. Step Up, Over, and Down									
5. Steps, Jumps, and Spins									
6. Nighttime Pink Barricade									
7. Gate Jumper									
8. Sidewalk-Benchrail-Grass									
9. Traffic Island Hopping									
10. Nighttime Wall to Wall									
11. Ten Steps Down									
12. Lateral Spot Jumps									
13. The Trees (With a Twist)									

Bicycle Dancer of Edinburgh

	Chemical PE to Grav PE	Chemical PE to Kinetic E	Kinetic E to Grav PE	Grav PE to Kinetic E	Dyn Unstable Equilibrium	Centripetal Acceleration	Rotation Vert Axis	Rotation Horiz Axis	Increase ↑ Decrease ↓ F
14. Rear Wheel Barhopping									
15. Up–Front…and Down									
16. Back (Wheel) and Forth									
17. Monorail									
18. 'Round the Corner Steps									
19. Up a Narrow Ramp									
20. Way Down 1									
21. Two-pylon Straddle									
22. More Cool Stunts									
23. Bike Shop—Copy Stop Hop									
24. Way Down 2									
25. Unicycle Up Over Up Down									
26. Way Down 3									

Name _____   Section _____   Date _____

CONCEPTUAL PHYSICS	Experiment

Thickness of a BB Pancake

Purpose
To determine the diameter of a ball bearing (BB) without directly measuring it

Apparatus
75 mL of BB shot
100-mL graduated cylinder
tray
ruler
micrometer

Discussion

This experiment distinguishes between **area** and **volume**, and sets the stage for the follow-up experiment, *Oleic Acid Pancake*, where you will estimate the size of molecules. To begin, consider the diagram of the eight wooden blocks arranged to form a single $2 \times 2 \times 2$-inch cube as shown to the right. What is the surface area of this cube? There are six faces to this cube and each face has a surface area of $2 \times 2 = 4$ square inches. The total surface area, therefore is 6×4 square inches = 24 square inches.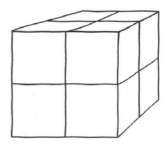

Any other arrangement of these cubes would result in a greater surface area. For example, laid out flat into a rectangular block, the surface area would be 28 square inches. If the blocks were stacked into an eight-block high tower, the surface area would be 34 square inches. Discuss with your lab partners how the greatest surface area is obtained when the blocks are spread out as much as possible.

Although different arrangements have different surface areas, the total **volume** remains the same. Notice that as the exposed surface area of the blocks changes, the total volume of all the blocks remains the same. Similarly, the volume of pancake batter is the same whether it is in the mixing bowl or spread onto a griddle. (Except that on a **hot** griddle the pancake gains surface area as cooking causes it to rise.)

How might you find the volume of a single pancake? The volume of a single pancake equals the surface area of one flat side multiplied by the thickness. If both the volume and the surface area are known, then the thickness can be calculated.

Volume = area × thickness

so simple rearrangement gives:

Thickness = $\dfrac{\text{volume}}{\text{area}}$

Instead of cubical blocks or pancake batter, consider a graduated cylinder that contains BBs. The space taken up by the BBs is easily read as volume on the side of the cylinder. If the BBs are poured into a tray, their volume remains the same. Can you think of a way to estimate the diameter (or thickness) of a single BB without measuring the BB itself? Try it in this experiment and see. It will be simply another step smaller to consider the size of molecules in the next experiment.

Procedure

Step 1: Use a graduated cylinder to measure the volume of the BBs. (Note that 1 mL = 1 cm³.)

Volume = _____ cm³

Step 2: Carefully spread the BBs out to make a compact layer one pellet thick on the tray. With a ruler, determine the area covered by the BBs. Describe your procedure and show your computations.

Area = _____ cm²

Step 3: Using the area and volume of the BBs, estimate the thickness (diameter) of a BB. Show your computations.

Estimated thickness = _____ cm

Step 4: Check your estimate by using a micrometer to measure the thickness (diameter) of a BB.

Measured thickness = _____ cm

Summing Up

1. How does your estimate compare with the measurement of the diameter of the BB? Calculate the percentage error (consult Appendix C on how to do this) between the estimated and measured thickness of the BB.

_____

_____

2. Oleic acid is an organic substance that is soluble in alcohol but insoluble in water. When a drop of oleic acid is placed in water, it usually spreads out over the water surface to create a *monolayer,* a layer that is one molecule thick. From your experience with BBs, describe a method for estimating the size of an oleic acid molecule.

_____

_____

CONCEPTUAL PHYSICS | **Experiment**

Oleic Acid Pancake

Purpose
To estimate the size of a single molecule of oleic acid

Apparatus
tray
water
chalk dust or lycopodium powder
eyedropper
oleic acid solution (5 mL oleic acid in 995 mL of ethanol)
10-mL graduated cylinder

Discussion
During this experiment you will estimate the **diameter** of a single molecule of oleic acid! The procedure for measuring the diameter of a molecule will be much the same as that of measuring the diameter of a BB in the previous experiment. The diameter is calculated by dividing the volume of the drop of oleic acid by the area of the **monolayer** film that is formed. The diameter of the molecule is the depth of the monolayer.

Volume = area × depth

$$\text{Depth} = \frac{\text{volume}}{\text{area}}$$

Procedure
Step 1: Pour water into the tray to a depth of about 1 cm. So that the acid film will show itself, spread chalk dust or lycopodium powder very lightly over the surface of the water. Use a minimum amount of powder because too much powder will prevent the oil from spreading optimally.

Step 2: Using the eyedropper, gently add a single drop of the oleic acid solution to the surface of the water. When the drop touches the water, the alcohol in it will dissolve in the water, but the oleic acid will not. The oleic acid spreads out to form a nearly circular patch on the water. Measure the diameter of the oleic acid patch in several places, and compute the average diameter of the circular patch.

Average diameter = _____ cm

The average radius is, of course, half the average diameter. Now compute the area of the circle ($A = 1/2\ \pi r^2$).

Area of circle = _____ cm^2

Step 3: Count the number of drops of oleic acid solution needed to occupy 1 mL (or 1 cm^3) in the graduated cylinder. Do this three times, and find the average number of drops in 1 cm^3 of solution.

Number of drops in 1 cm^3 = _____

Divide 1 cm^3 by the number of drops in 1 cm^3 to determine the volume of a single drop.

Volume of single drop = _____ cm^3

Step 4: The volume of the oleic acid alone in the circular film is much less than the volume of a single drop of the solution. The concentration of oleic acid in the solution is 5 mL per liter of solution. Every cubic centimeter of the solution thus contains only $\frac{5}{1000}$ cm^3, or 0.005 cm^3, of oleic acid. The volume of oleic acid in one drop is thus 0.005 of the volume of one drop. Multiply the volume of a drop by 0.005 to find the volume of oleic acid in the drop. This is the volume of the layer of acid in the tray.

Volume of Oleic acid = _____ cm^3

Step 5: Estimate the diameter of an oleic acid molecule by dividing the volume of oleic acid by the area of the circle.

Diameter = _____ cm

The diameter of an oleic acid molecule as obtained by this method is good, but not precise. This is because an oleic acid molecule is not spherical, but rather elongated like a hot dog. One end is attracted to water, and the other end points away from the water surface. The molecules stand up like people in a puddle! So the estimated diameter is actually the estimated length of the short side of an oleic acid molecule.

Summing Up

1. What is meant by a **monolayer**?

_____

2. Why is it necessary to dilute the oleic acid for this experiment? Why alcohol?

_____

3. The shape of an oleic acid molecule is more like that of a hot dog than a sphere. Furthermore, one end is attracted to water (**hydrophilic**) so that the molecule stands up on the surface of water. Assume an oleic molecule is 10 times longer than it is wide. Then estimate the volume of one oleic acid molecule.

_____

_____

CONCEPTUAL PHYSICS | **Experiment**

Totally Stressed Out

Purpose
To find the relationship between the stressing force acting on a spring and the subsequent stretching of the spring: Hooke's Law

Apparatus
table clamp or support base   support rod
crossbar   collar hook
mass hanger   meterstick
clamp (meterstick, 3-finger, buret, or equivalent for holding the meterstick in place)
slotted masses (variety, including one or more 50-g, 100-g, and 200-g)
spring (with an appropriate force constant for the masses available)
sheet of graph paper

Going Further
rubber band (#64, for example)

Discussion
An understanding of the behavior of a spring loaded with masses might seem to have limited value. In reality, all objects act like springs to some extent: even the strongest solid objects can be thought of as very stiff springs. Therefore, you are really exploring the behavior of anything under stress. You'll use springs in this lab only because their behavior is easier to observe and analyze than other objects.

The scientist who is given credit for first discovering this relationship was Robert Hooke, an industrious English scientist who was a contemporary of Isaac Newton. Newton saw Hooke as a rival, and Hooke claimed that Newton plagiarized some of his work. Hooke died when Newton was President of The Royal Society, to which England's most prestigious scientists belonged. Hooke's Royal Society portrait did not survive Newton's term of office, and the location of Hooke's final burial is not known. But Hooke's law of elasticity survives, as you will find in this experiment.

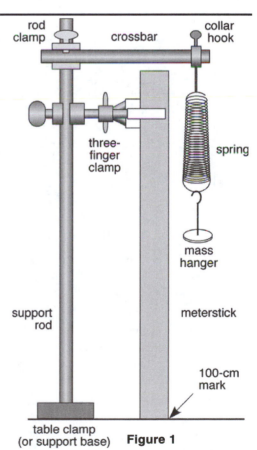

Figure 1

Procedure
Step 1: Arrange the apparatus in a manner similar to what is shown in Figure 1.

a. The meterstick is oriented with the 0-cm mark on top and the 100-cm mark on bottom.

b. Adequate clearance is left below the spring so that the spring can stretch when slotted masses are added to its load.

c. The meterstick is close to the mass hanger.

Step 2: If the coils of the spring touch each other even after the mass hanger is attached, the spring is under tension. For best results, add enough slotted mass so the coils no longer touch.

Step 3: Note the position of the mass hanger when the coils of the spring do not touch. (Some springs are not under tension; their coils are free of each other when the mass hanger is first attached.) Observe the location of the mass hanger using the nearby vertical meterstick. Record the value on the data table below.

Step 4: Add slotted mass (perhaps 50 or 100 grams) to generate the next data point. Record the added mass and the new location of the hanger.

Step 5: Continue to generate new data points by adding mass in 50- or 100-gram increments. Do not exceed 500 grams or whatever limit your instructor announces.

Remember that the total mass value for each data point is the mass added *after* the original, "unloaded" position was recorded. Leave the three remaining columns of the data table blank while collecting data.

Depending on the number of data points you collect, some rows on the data table may be left blank.

Data Table

Total Mass (g)	Position (cm)	Total Mass (kg)	Stretch x (m)	Load Weight F (N)
0		0	0	0

Step 6: Determine and record the mass in kilograms of all added slotted mass values by using the correct conversion factor. There are 1000 grams in 1 kilogram. For example, 100 grams is 0.100 kilograms.

Step 7: Calculate and record the stretch of the spring by subtracting each position data point from the original position value. Convert the value to meters by using the correct conversion factor. There are 100 centimeters in 1 meter. For example, if the original position was 47.3 cm and a stretched position was 83.4 cm, the stretch would be 83.4 cm – 47.3 cm = 36.1 cm and therefore 0.361 m.

Step 8: Calculate and record the load weight of each added mass value by using the equation $W = mg$. For example, the load weight of 0.100 kg is found by $W = mg = 0.100 \text{ kg} \times 9.8 \text{ m/s}^2 = 0.980 \text{ N}$. In the space below, show the calculation of the weight of the largest added mass you used (which should be 0.400 kg or greater).

_____

_____

Step 9: On your graph paper, create a graph titled "Load vs. Stretch." (If you are unsure of which quantity to place on the vertical axis and which to place on the horizontal axis, see Appendix E on graphing at the back of this lab manual.) Label the axes completely and correctly, and scale the graph appropriately to maximize its area on your graph paper. Plot the load vs. stretch data from your data table.

Step 10: Obtain an object of unknown mass from your instructor. Suspend the object from your spring in its original, "unloaded" state. (This will include the hanger and any mass needed to separate the coils as discussed in Step 2.) Record the position of the hanger when the spring is loaded with the unknown object.

_____

Summing Up

1. Notice that the plotted graph forms a reasonably linear (straight-line) plot.
    a. How can you determine the slope of the plot?

_____

_____

_____

    b. What is the numerical value of the slope of the plot?

_____

_____

    c. What are the units of the slope in this case?

_____

2. Which is a better interpretation of the meaning of the numerical value slope of the plot?

    ____ The number of newtons of load force required to stretch the spring one meter.

    ____ The number of meters that the spring would stretch for one newton of force.

3. The slope of the plot is the force constant (or spring constant) of the spring, denoted with the symbol k. This is not a universal constant. Nor is it a value that applies to all springs. The force constant specifies the stiffness of the specific spring whose load and stretch values were plotted. Stiffer springs have greater force constants; weaker springs have lesser force constants. What is the force constant of the spring you observed? _Don't forget to write the units!_

    k = _____

4. What is the relationship between loading force (F), stretch (x), and force constant (k)? Write an equation that includes all three quantities in the space below.

5. a. How can you determine the unknown mass of the object you were given?

_____

_____

_____

_____

_____

    b. What is the mass of the object as determined by your process?

_____

    c. What is the actual mass of the object as given by your instructor?

_____

    d. What is the percent error in your determination of the unknown's mass? (If you're not sure how to calculate percent error, see Appendix C in the back of the manual.)

_____

_____

6. Use the plot of your spring to answer the following questions.

    a. What is the weight of an object that would stretch your spring 0.27 m when suspended from it?

_____

    b. To what extent would an object weighing 3.25 N stretch your spring if suspended from it?

_____

7. Where would the plot of a *stiffer* spring lie on the graph? Draw and label such a plot on your graph and describe its slope in the space below.

_____

8. Where would the plot of a *weaker* spring lie on the graph? Draw and label such a plot on your graph and describe its slope in the space below.

_____

Going further

Replace your spring with a strong rubber band. Carefully repeat the procedure used to generate data points for the spring. Plot those data points on your graph. How does the behavior of the rubber band compare with the behavior of the spring?

_____

_____

CONCEPTUAL PHYSICS	Tech Lab

Spring to Another World

Purpose

To use a simulation of masses and springs to determine force constants, a mass value, and the gravitational acceleration of an unknown planet

Apparatus

computer   PhET simulation: "Masses and Springs" (at http://phet.colorado.edu)
additional sheet of writing paper

Discussion

Seventeenth-century English scientist Robert Hooke is credited with the discovery that the force exerted by a spring is directly proportional to the length it is stretched or compressed. This simulation has been programmed to obey Hooke's law. It will allow you to practice good lab technique to solve a few simple puzzles.

Procedure

SETUP

Step 1: Turn on the computer and let it complete its start-up process.

Step 2: Open the PhET simulation, "Masses and Springs." If you're not sure how to do this, ask your instructor for assistance.

Step 3: When the simulation opens, the screen should resemble Figure 1 below.

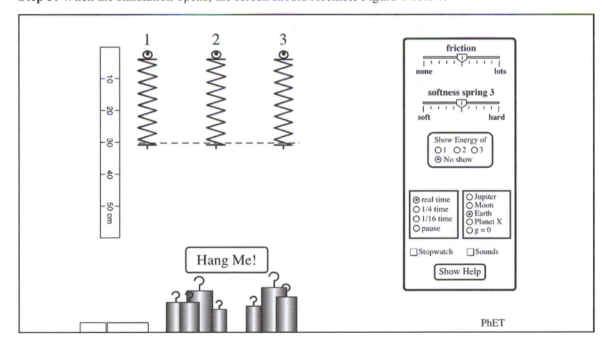

Figure 1. Masses and Springs

PART A: DETERMINATION OF A FORCE CONSTANT

Step 1: If it has not already been done, select Earth in the on-screen control panel.

Step 2: Click and drag to move the on-screen horizontal dashed line so that it is aligned with the bottom of the three springs.

Step 3: Click and drag to move the on-screen ruler so that its top (0 cm) is aligned with the dashed line.

Step 4: Click and drag to attach a 100-gram hooked mass to Spring 1. Determine the load force (F) of the 100-gram mass by converting grams to kilograms, then use $F = mg$.

_____

_____

Step 5: Carefully record the amount of stretch (x) that the spring experiences when loaded with the 100-gram mass.

_____

Step 6: Rearrange Hooke's Law, $F = kx$, solving for k. Then determine the force constant (k) of Spring 1 using the force from Step 7 and the stretch from Step 8.

_____

_____

PART B: THE RED MATTER

Step 1: On a separate sheet of paper, describe a method to determine the mass of the red cylinder.

Step 2: Also on that sheet, record the data and any calculations needed to determine the mass of the red cylinder. Organize your data neatly and show calculations completely.

Step 3: Record the mass of the red cylinder here: m = _____.

PART C: GRAVITATIONAL ACCELERATION ON PLANET X

Suppose you were going to travel from Earth to Planet X. You can take Spring 1 and the gold cylinder with you on your voyage.

Step 1: On a separate sheet of paper, describe a method to determine the gravitational acceleration on Planet X using Spring 1 and the gold cylinder. You may conduct experiments on both worlds, and you may use knowledge gained in previous steps. But you may not use any other masses or springs.

Step 2: Also on that sheet, record the data and any calculations needed to determine the gravitational acceleration on Planet X. Organize your data neatly and show calculations completely.

Step 3: Record the gravitational acceleration of Planet X here: g = _____.

PART D: THE RANGE OF FORCE CONSTANTS FOR SPRING 3

Return to Earth (via the on-screen planet selection). Notice that there is an on-screen slide switch that can be used to adjust the force constant of Spring 3.

Step 1: On a separate sheet of paper, describe a method to determine the lowest and highest force constant values that Spring 3 can be set to.

Step 2: Also on that sheet, record the data and any calculations needed to determine the extreme force constant values of Spring 3. Organize your data neatly and show calculations completely.

Step 3: Record the *lowest* Spring 3 force constant value here: k_L = _____.

Step 4: Record the *highest* Spring 3 force constant value here: k_H = _____.

CONCEPTUAL PHYSICS	**Activity**

Eureka!

Purpose
To investigate the displacement of water by immersed objects

Apparatus
35-mm film canisters (or small, plastic, carry-on bottles for liquids) with string attached
ballast material (nuts, BBs, sand, etc.)   balance
500 mL graduated cylinders   access to masking tape

Discussion
Archimedes is much remembered for his clever discovery of the method for determining the volume of irregularly shaped objects—like himself! He discovered this while in the public baths of Athens, pondering a problem given to him by the king—how to determine whether a particular gold crown was really 100% gold or not. His discovery is based on a very simple idea that many people don't fully understand. If perchance you are one of them, after doing this activity, you won't be!

Procedure
Step 1: Your instructor will give you two film canisters, each with contents having different masses, and each with a piece of string attached. Find the mass of each canister.

Mass of lighter canister, m = _____ g

Mass of heavier canister, m = _____ g

Step 2: So that you can easily mark the water level in a graduated cylinder, attach a vertical strip of masking tape along its outer side. Fill the cylinder about three-quarters full of water. Mark the water level on the tape. Submerge the lighter canister and mark the new water level on the tape. Remove the canister.

Step 3: Predict how the water level for the heavier canister will compare when it is submerged.

_____

Step 4: Try it and see! Was your prediction correct?

_____

Summing Up
Describe and explain your observations.

_____

_____

The most exciting phrase to hear in science, the one that heralds new discoveries, is not 'Eureka!' but 'That's funny. . . .'

Isaac Asimov

CONCEPTUAL PHYSICS	Activity

Sink or Swim

Purpose
To observe the effect of density on various objects placed in water

Apparatus
equally massive blocks of lead, wood, and Styrofoam
at least two aluminum cans of soda pop—both diet and regular
aquarium tank or sink two-thirds filled with water
egg
wide-mouth graduated cylinder
bowl
salt
spoon
balance

Discussion
Why can some people float in water while others can't? The answer has to do with density. This activity should increase your understanding of the role of density in floating.

Procedure
Step 1: Balance equal masses of lead and wood, using a double-beam balance. Repeat, using an equal mass of Styrofoam.

How do the **volumes** compare? How do the **densities** therefore compare?

_____

_____

Step 2: Try floating cans of soda pop in the aquarium/sink.

a.   Which ones float? Sink?

_____

_____

b.   How does the density of the different kinds of soda pop compare with the density of tap water?

_____

_____

c. Hypothesize how the relative densities relate to sugar content.

_____

_____

Step 3: Use a balance to measure the mass of an egg. Using a wide-mouth graduated cylinder, carefully determine the volume of the egg by measuring the volume of water it displaces when it is slowly (gently!) lowered into the graduated cylinder. Calculate its density: $d = m/V$.

Density = _____

Step 4: Now try to float the egg in a bowl of water. Does it float? If not, dissolve enough salt in the water until the egg floats.

a. How does the density of an egg compare with that of tap water?

_____

_____

b. With saltwater?

_____

_____

Summing Up

1. Does adding salt to the water make the water less dense or more dense? How?

_____

_____

_____

_____

2. Why do some people find it difficult to float while others don't? What evidence can you cite for the notion that it is easier to float in salt water?

_____

_____

_____

_____

CONCEPTUAL PHYSICS | **Activity**

Boat Float

Purpose
To investigate Archimedes' Principle and the principle of flotation

Apparatus
spring scale | 100-g mass
string and masking tape | balance
600-mL beaker | rock or hook mass
clear container or 3-gallon bucket | 500-mL graduated cylinder
chunk of wood | water
toy boat capable of carrying a 1200-gram | modeling clay
cargo, or 9-inch aluminum cake pan | 3 lead masses or lead fishing weights

Discussion

An object submerged in water takes up space and pushes water out of the way—the water is *displaced.* Interestingly enough, the water that is pushed out of the way pushes back on the submerged object. For example, if the object pushes a volume of water with a weight of 100 N out of its way, then the water reacts by pushing back on the object with a force of 100 N—Newton's third law. The object is *buoyed* upward with a force of 100 N. This is summed up in Archimedes' principle, which states that the *buoyant force* that acts on any completely or partially submerged object *is equal to the weight of the fluid the object displaces.*

Procedure
Step 1: Use a spring scale to determine the weight of an object (rock or hook mass) that is first out of water and then under water. The difference in weights is the buoyant force. Record.

Weight of object out of water = _____

Weight of object in water = _____

Buoyant force on object = _____

Step 2: Devise a method to find the volume of water displaced by the object. Record the volume of water displaced. Compute the mass and weight of this water. (Remember, 1 mL of water has a mass of 1 g and a weight of 0.01 N.)

Volume of water displaced = _____

Mass of water displaced = _____

Weight of water displaced = _____

How does the buoyant force on the submerged object compare with the weight of the water displaced?

_____

_____

NOTE: To simplify calculations for the remainder of this activity, measure and determine *masses,* without the need of finding their equivalent *weights* ($W = mg$). Keep in mind, however, that an object floats because of a buoyant *force.* This force is due to the *weight* of the water displaced.

Step 3: Use a balance to measure the mass of a piece of wood, and record the mass in the data table. Measure the volume of water displaced when the wood floats. Record the volume and mass of water displaced in the data table.

a. What is the relation between the buoyant force on any floating object and the weight of the object?

_____

_____

b. How does the mass of the wood compare with the mass of the water displaced?

_____

_____

c. How does the buoyant force on the wood compare with the weight of water displaced?

_____

_____

Step 4: Add a 100-g mass to the wood so that the wood displaces more water but still floats. Measure the volume of water displaced, and calculate its mass. Record them in the data table.

How does the buoyant force on the wood with its 100-g load compare with the weight of water displaced?

_____

_____

Step 5: Roll the clay into a ball and find its mass. Measure the volume of water it displaces after it sinks to the bottom of a graduated cylinder. Calculate the mass of water displaced. Record all volumes and masses in the data table.

a. How does the mass of water displaced by the clay compare with the mass of the clay out of the water?

_____

_____

b. Is the buoyant force on the submerged clay greater than, equal to, or less than its weight out of the water? What is your evidence?

_____

_____

Step 6: Retrieve the clay from the bottom, and mold it into a shape that allows it to float. Sketch or describe this shape. Measure the volume of water displaced by the floating clay. Calculate the mass of the water, and record in the data table.

Data Table

Object	Mass (g)	Volume of Water Displaced (mL)	Mass of Water Displaced (g)
Wood			
Wood + 50-g Mass			
Clay Ball			
Floating Clay			

Summing Up

1. Does the clay displace more, less, or the same amount of water when it floats as it did when it sank?

_____

2. Is the buoyant force on the floating clay greater than, equal to, or less than the weight of the clay?

_____

3. What can you conclude about the weight of an object and the weight of water displaced by the object when it floats?

_____

4. Is the buoyant force on the clay ball greater when it is submerged near the bottom of the container or when it is submerged near the surface? What is your evidence?

_____

_____

5. Is the pressure that the water exerts on the clay ball greater near the bottom of the container than when submerged near the surface? Exaggerate the depth involved and cite expected evidence.

_____

_____

6. Why are your last two answers different?

_____

Going Further

1. Suppose you are on a ship in a canal lock. If you throw a ton of lead bricks overboard from the ship into the canal lock, will the water level in the canal lock go up, down, or stay the same? Write down your prediction *before* you proceed to Step 8.

   Prediction for water level in canal lock: _____

2. Float a toy boat loaded with lead "cargo" in a relatively deep container filled with water (deeper than the height of the lead masses). For observable results, the size of the container should be just slightly bigger than the boat. Mark and label the water levels on masking tape placed on the container and on the sides of the boat. Remove the masses from the boat and put them in the water. Mark and label the new water levels.

3. What happens to the water level *on the side of the boat* when you remove the cargo? What does this say about the amount of cargo carried by ships that float high in the water?

   _____

4. What happens to the water level *in the container* when you place the cargo in the water? Explain why this happens.

   _____

   _____

5. Similarly, what happens to the water level in the canal lock when the bricks are thrown overboard?

   _____

6. Suppose the freighter is carrying a cargo of Styrofoam instead of bricks. What happens to the water level in the canal lock if the Styrofoam (which floats in water) is thrown overboard?

   _____

7. When a ship is launched at a shipyard, what happens to the sea level all over the world—no matter how imperceptibly?

   _____

8. When a ship in the harbor launches a little rowboat from the dock, what happens to the sea level all over the world—no matter how imperceptibly?

   _____

9. Cite evidence to support your (different?) answers to 7 and 8.

   _____

10. One of the most fascinating applications of a floating object displacing a weight of water equal to its own weight is the Falkirk Wheel, featured in the photo opener to Chapter 13, and Figure 13.19 in your textbook. Explain why only a low-power motor can lift heavy ships in this way.

   _____

   _____

   _____

CONCEPTUAL PHYSICS   **Experiment**

Chapter 14: Gases   Force, Area, and Pressure

Tire Pressure and 18-Wheelers

Purpose
To investigate the difference between pressure and force

Apparatus
automobile   tire pressure gauge
graph paper   owner's manual for vehicle (optional)

Discussion
People commonly confuse *force* and *pressure*. Tire manufacturers add to this confusion by saying, "Inflate to 45 pounds," when they really mean, "45 pounds *per* square inch." The pressure on your feet is painfully more when you stand on your toes than when you stand on your whole feet, even though the force of gravity (your weight) is the same. Pressure depends on how a force is distributed. Pressure increases as area decreases, and decreases as area increases.

$$\text{Pressure} = \frac{\text{force}}{\text{area}}$$

$$\text{Force} = \text{pressure} \times \text{area}$$

Ever wonder why trucks that carry heavy loads have so many tires?
Count the tires on the trucks you see on the highway. Some commonly have 18! Why?

Procedure
Step 1: Position a piece of graph paper directly in front of the front tire of an automobile. Roll the automobile onto the paper.

Step 2: Use a tire gauge to measure the pressure in each tire in pounds per square inch. Record the pressures of each tire in the data table. Trace the outline of the tire where it makes contact with the graph paper. Roll the vehicle off the paper. Calculate the area inside the trace in square inches or square cm. Record your data in the data table.

Step 3: Study the tread pattern of the tire. Compare this with the tire mark on the paper. Note that only the rubber in the tread actually presses against the road. The gaps in the tread do not support the vehicle. Estimate what fraction of the tire's contact area is tread and record it in the data table. The area actually in contact with the road is the area inside your trace multiplied by the fraction of tread. Repeat for all four tires.

Step 4: Compute the force each tire exerts against the road by multiplying the pressure of the tire times its area of contact with the road. Record your computations in the data table.

Data Table

Tire	Area Traced on Paper (in²)	Estimate of % Tread	Area of Contact	Pressure (lb/in²)	Force (lb)
Right Front					
Left Front					
Right Rear					
Left Rear					

Step 5: Compute the weight of the vehicle by adding the forces from each tire.

Computed weight = _____ lb

Is this not quite remarkable—that the weight of a vehicle is equal to the air pressure in its tires multiplied by the area of tire contact? (See if your Uncle Harry knows this!)

Step 6: Ascertain the weight of the vehicle from another source, such as the owner's manual or the local dealer. Sometimes this information is stamped on a plate on the inside of a doorjamb.

Known weight = _____ lb

Summing Up

1. How does your computed value of vehicle weight compare with the stated value? What might account for a difference in the actual weight and the weight stated in the owner's manual?

_____

2. When you inflate tires to a higher pressure, what happens to the contact area of the tire against the road surface?

_____

3. A flat tire will read zero on a pressure gauge, when actually air at atmospheric pressure is inside. So a tire gauge is calibrated to measure the air pressure over and above atmospheric pressure. Consider the difference between **gauge pressure** and **atmospheric pressure**. Atmospheric pressure is normally 14.7 lb/in² or 10^5 N/m² outside the tire. Gauge pressure is the amount of pressure inside the tire over and above atmospheric pressure. So what is the **total** pressure inside the front tire?

_____

4. Why was the atmospheric pressure (14.7 lb/in²) **not** added to the pressure of the tire gauge when computing the weight of the vehicle?

_____

5. Why do trucks that carry heavy loads have so many wheels—often 18?

_____

CONCEPTUAL PHYSICS	**Demonstration**

Dance of the Molecules

Purpose
To observe the difference between hot water and cold water—on the *molecular* level

Apparatus
access to cold water and hot water
2 empty baby food jars or small beakers
food coloring (2 colors preferred)

Going Further
2 index cards (or playing cards)
hot-surface hand protection (Ove-Glove or equivalent)

Discussion
The difference between hot things and cold things is referred to as *temperature.* The temperature of an object depends on the kinetic energy of the random motions of its molecules, as noted in Figure 15.3 in your text. Without a thermometer we most often can't distinguish a hot thing from a cold thing. Consider a glass of hot water and a glass of cold water side by side. The water molecules in each glass don't appear to be moving differently from one another. But they are. Follow the steps below to *see* the difference.

Procedure
Step 1: Fill one jar with cold water and the other with hot water. Fill them to the brim.

Step 2: Spend a minute or two making a prediction. What will happen if a few drops of food coloring are added to each jar of water? Record your prediction using words and pictures. (Let the water stand while you record your prediction. This will allow swirling currents and turbulence in the water to diminish.)

_____

_____

_____

Step 3: Carefully add just a drop or two of food coloring to each jar. Watch the coloring spread out for 1 minute before recording your observations, taking care not to disturb the water in any way. Record your observations using words and pictures.

_____

_____

_____

Summing Up

1. How do your observations support the principle that the molecules in hot objects move faster than the molecules in cold objects?

_____

_____

2. Suppose you let the experiment continue for several hours. What would the jars of water and food coloring look like afterward?

_____

_____

3. Consider a fragrant candle burning in the corner of one side of a large room. It would be possible to smell the fragrance on the other side of the room after a short period of time. Use your observations to explain how this is so.

_____

_____

4. Would the fragrance reach you more quickly, more slowly, or in the same amount of time if the air in the room were colder?

_____

_____

Going Further

If you are using glass baby food jars, you can extend the activity.

Step 1: Carefully place an index card on top of the hot water jar. The bootom of the card should get wet with hot water. Push the card down if necessary. Leave the card in place. Protect your hand with the Ove-Glove (or equivalent). With the protected hand, hold the hot water jar over a sink and quickly invert it. If everything works as planned, the hot water will remain in the jar, seemingly held in place by the index card! *Do not attempt to hold the card while inverting the hot water!* This trick may require some practice.

Step 2: Carefully set the inverted hot water on top of the open mouth of the cold water jar.

Step 3: Carefully remove the card. Observe and record the result.

_____

_____

Step 4: Carefully insert a dry card between the hot and cold water jars.

Step 5: Carefully invert the two-jar assembly.

Step 6: Carefully remove the card. Observe and record the result.

_____

_____

CONCEPTUAL PHYSICS   **Tech Lab**

Bouncing Off the Walls

Purpose
To control and observe the behavior of gas particles (atoms or molecules) as modeled in a simulation to investigate properties of gas such as temperature and pressure

Apparatus
computer   PhET simulation: "Gas Properties" (available at http://phet.colorado.edu)

Discussion
Kinetic molecular theory explains the large-scale characteristics of gases in terms of the behavior of the atoms and molecules that make up the gas. The "Gas Properties" simulation lets you see the individual particles in motion. It gives you control of a chamber of gas and lets you see the effects of the changes you make.

Setup
Step 1: Start the computer and log in. Open the PhET simulation, "Gas Properties."

Step 2: In the on-screen control panel, click the "Measurement Tools" button.

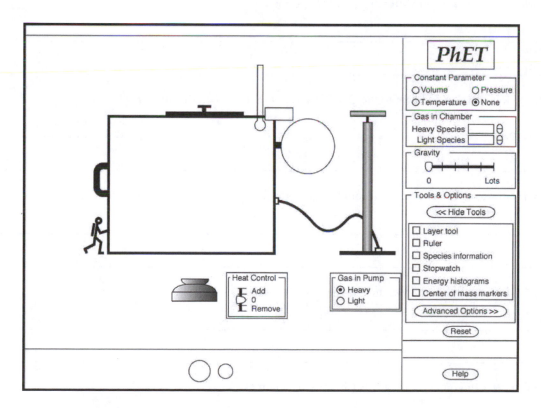

Step 3: Add labels to the figure above for each item listed below.

Thermometer   *Pressure gauge*   *Heat source/sink*   *Gas pump*

Chamber lid   *Play/pause*   *"Scubie" (the volume adjuster)*

Procedure

PART A: SIMULATION MECHANICS

Remember that any time you need to, you can use the on-screen "Reset" button to return to the initial setup.

Step 1: Determine two distinct methods by which you can add particles to the chamber.

a. Method 1: Use the pump. How do you manipulate the pump handle to get the *greatest* number of particles into the chamber in *one* stroke?

_____

_____

b. Describe Method 2: How can you *precisely* control the number of particles injected into the chamber? (Method 2 does not involve direct use of the pump.)

_____

_____

Step 2: How can you release particles from the chamber (*without* breaking the chamber)?

a. Method 1:

_____

_____

b. Method 2 (*completely* different from Method 1):

_____

_____

Step 3: How can you add heat to the gas? How does the simulation illustrate this?

_____

_____

Step 4: How can you remove heat from the gas? How does the simulation illustrate this?

_____

_____

Step 5: How can you compress the gas (decrease its volume)?

_____

_____

Step 6: How can you expand the gas (increase its volume)?

_____

_____

PART B: THE NATURE OF THE IDEAL GAS LAW

Step 1: What happens to the pressure in the chamber if you heat the gas?

_____

Step 2: What happens to the temperature in the chamber if you compress the gas?

_____

Step 3: Locate the "Constant Parameter" section of the on-screen control panel. Lock the temperature. Use "Scubie" to slowly compress the gas. What happens to the temperature, and what action is taken (by the simulation) to maintain constant temperature?

_____

_____

Step 4: Lock the pressure. (Doing so releases the lock on temperature.) Add heat. What happens to the pressure, and what action is taken (by the simulation) to maintain constant pressure?

_____

_____

PART C: ALL SPECIES GREAT AND SMALL
Create a chamber in which there is a mix of 50 light and 50 heavy gas particles. In the on-screen control panel, activate "Species Information."

Step 1: Consider the species information and the meaning of temperature.

a. Which species—if either—has the greater average speed?

_____

b. Does the temperature in the chamber reflect the average speed, momentum, potential energy, or kinetic energy of the particles? And how does this explain your finding in Step 1. a.?

_____

_____

_____

Step 2: Reset the chamber to have about 100 heavy particles. The temperature should be 300 K. Click the on-screen button to activate the "Center of mass markers."

a. Consider the statement, "The average *speed* of air molecules in a room may be over 1000 mph (400 m/s), while their average *velocity* is approximately zero." The particles in the simulation's chamber are modeling the air molecules in a room. Cite evidence from the simulation to confirm or reject the quoted statement.

_____

_____

_____

_____

Bouncing Off the Walls

b. What is the name of the condition when the average velocity of atmospheric molecules around you is not zero? (*Hint:* It's a common four-letter word starting with the letter "W.")

_____

Going Further

1. Under what specific condition will the lid be blown off the chamber? Is it possible to blow the lid with just one particle in the chamber?

_____

_____

_____

2. Is it possible to achieve a six-figure temperature? If so, how? If not, what's the highest temperature you could achieve?

_____

_____

_____

3. As of 2010, low-temperature physicists have not been able to cool anything to a temperature of absolute zero (0° K).

    a. Cool a simulated 50/50 sample (50 light and 50 heavy particles) gas sample to absolute zero.

    b. When the temperature hits absolute zero, are the particles shown to be at rest?

_____

_____

4. Use Scubie and the ice ("Heat Control: Remove") to compress and cool a 50/50 sample to the smallest volume possible and to absolute zero. Now use Scubie to rapidly expand the volume of the chamber all the way out. Describe what happens. And which species wins the race to the far side of the chamber (where Scubie is)?

_____

_____

_____

_____

_____

_____

| **CONCEPTUAL PHYSICS** | **Experiment** |

Chapter 15: Heat, Temperature, and Expansion   Specific Heat Capacity

Temperature Mix

Purpose
To predict the final temperature of a mixture of cups of water at different temperatures

Apparatus
3 Styrofoam cups
liter container with a wide mouth
thermometer (Celsius)
pail of cold water
pail of hot water

Discussion
If you mix a pail of cold water with a pail of hot water, the final temperature of the mixture will be between the two initial temperatures. What information would you need to predict the final temperature? You'll begin with the simplest case of mixing *equal* masses of hot and cold water.

Procedure
Step 1: Begin by marking your three Styrofoam cups equally at about the three-quarter mark. You can do this by pouring water from one container to the next and mark the levels along the inside of each cup.

Step 2: Fill the first cup to the mark with hot water from the pail, and fill the second cup with cold water to the same level. Measure and record the temperature of both cups of water.

  Temperature of cold water = _____

  Temperature of warm water = _____

Step 3: Predict the temperature of the water when the two cups are combined. Then pour the two cups of water into the liter container, stir the mixture slightly, and record its temperature.

  Predicted temperature = _____

  Actual temperature of water = _____

If there was a difference between your prediction and your observation, what may have caused it?

Pour the mixture into the sink or waste pail. (Don't be a klutz and pour it back into either of the pails of cold or hot water!) Now you'll investigate what happens when *unequal* amounts of hot and cold water are combined.

Step 4: Fill one cup to its mark with cold water from the pail. Fill the other two cups to their marks with hot water from the other pail. Measure and record their temperatures. Predict the temperature of the water when the three cups are combined. Then pour the three cups of water into the liter container, stir the mixture slightly, and record its temperature.

    Predicted temperature = _____

    Actual temperature of water = _____

Pour the mixture into the sink or waste pail. Again, do **not** pour it back into either of the pails of cold or hot water!

a. How did your observation compare with your prediction?

_____

b. Which of the water samples (cold or hot) changed more when it became part of the mixture? In terms of energy conservation, suggest a reason for why this happened.

_____

Step 5: Fill two cups to their marks with cold water from the pail. Fill the third cup to its marks with hot water from the other pail. Measure and record their temperatures. Predict the temperature of the water when the three cups are combined. Then pour the three cups of water into the liter container, stir the mixture slightly, and record the temperature.

    Predicted temperature = _____

    Actual temperature of water = _____

Pour the mixture into the sink or waste pail. (By now, you and your lab partners won't alter the source temperatures by pouring waste water back into either of the pails of cold or hot water.)

a. How did your observation compare with your prediction?

_____

b. Which of the water samples (cold or hot) changed more when it became part of the mixture? Suggest a reason for why this happened.

_____

_____

Summing Up

1. What determines whether the equilibrium temperature of a mixture of two amounts of water will be closer to the initially cooler or warmer water?

_____

2. How does the formula $Q = mc\Delta T$ apply here?

_____

CONCEPTUAL PHYSICS	Experiment

Spiked Water

Purpose
To determine which is better able to increase the temperature of a quantity of water: a mass of hot nails or the same mass of equally hot water

Apparatus
equal arm balance (Harvard trip balance or equivalent)
4 large insulated cups
bundle of short, stubby nails tied together with string
thermometer (Celsius)
hot and cold water
paper towels

Discussion
Suppose you have cold feet when you go to bed, and you want something to keep your feet warm throughout the night. Would you prefer to have a bottle filled with hot water, or one filled with an equal mass of nails at the same temperature as the water? The one that can store more thermal energy will do the better job. But which one is it? In this experiment, you'll find out.

Procedure
Consider the equal masses of nails and water, both warmed to the same temperature. Which one will do a better job of heating a sample of cold water? Make a prediction before performing the experiment.

Step 1: Fill two cups 1/3 full of *cold* water. Place one cup on each pan of the balance to make sure they contain equal masses of water, as shown in Figure 1. Add water to the lighter side until they do. Make sure your bundle of nails can be completely submerged in the cold water.

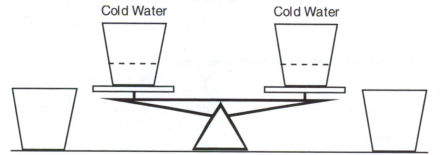

Cold Water   Cold Water

Figure 1

Step 2: Dry the nails.

Step 3: Place a large empty cup on each pan of a beam balance. Place the bundle of nails into one of the cups. Add **hot** water to the other cup until it balances the cup of nails, as shown in Figure 2. When the two cups are balanced, the mass in each cup is the same.

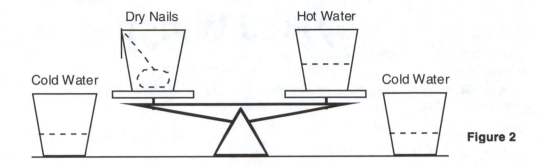

Figure 2

Step 4: Lower the bundle of nails into the hot water as shown in Figure 3. Be sure that the nails are completely submerged by the hot water. Allow the nails and the water to reach thermal equilibrium. (This will take a minute or two.)

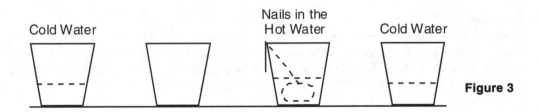

Figure 3

Step 5: Once the nails and hot water have come to thermal equilibrium, take the nails out of the hot water and put them in one of the cups of cold water. Pour the remaining hot water into the other cup of cold water. See Figure 4.

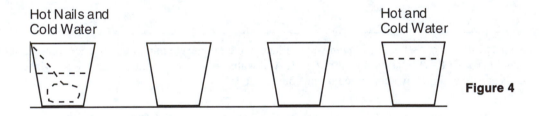

Figure 4

Step 6: Measure and record the final temperature of the mixed water.

T_W = _____ °C

Step 7: When the temperature of the nail-water mix stops rising, measure and record it.

T_N = _____ °C

Summing Up

1. Which was hotter before being put into the cold water, the nails or the hot water the nails were soaking in? How do you know?

_____

_____

2. Which was more effective in raising the temperature of the cold water, the hot nails or the hot water?

_____

_____

3. Suppose you have cold feet when you go to bed, and you want something to warm your feet throughout the night. Would you prefer to have a bottle filled with hot water, or one filled with an equal mass of nails at the same temperature as the water? Relate your answer to your findings in this experiment.

_____

_____

4. A student who conducted this experiment suggested that the temperature of the nail-water mix rose more than it should have because some water clung to the nails when they were transferred from the hot water to the cold water. Another student says that this caused the temperature of the nail-water mix to rise less than it should have if no water clung to the nails during the transfer. Who do you believe and why?

_____

_____

_____

_____

Spiked Water

Judge a man by his questions rather than by his answers.

Voltaire

CONCEPTUAL PHYSICS **Experiment**

Canned Heat: Heating Up

Purpose
To compare the ability of different surfaces to absorb thermal radiation

Apparatus

heat lamp and base	radiation cans: silver, black, and white
thermometer	access to cold tap water
paper towel	graph paper

Discussion
Does the color of a surface make a difference in how well it absorbs thermal radiation? If so, how? The answer to these questions could help you decide what to wear on a hot, sunny day and what color to paint your house if you live in a hot, sunny climate. In this experiment, you will compare the thermal absorption ability of three surfaces: silver, black, and white. You'll do this by filling cans with these surfaces with water, then exposing the cans to heat lamps. You'll measure the temperature of the water in each of the cans while they're being exposed to the heat and see if there's a difference in the rate at which the temperatures increase.

Pre-lab Questions
1. In which of the three cans do you think the water will heat up at the fastest rate?

2. In which of the three cans do you think the water will heat up at the slowest rate?

Procedure
Step 1: Arrange the apparatus so that the heat lamp will shine equally on all three cans. The cans should be about 1 foot in front of the lamp. Do not turn the light on yet.

Step 2: Fill the cans with cold water and wipe up any spills. Quickly measure the initial temperature of the water in each of the cans and record the numbers in the data table.

Step 3: Turn on the lamp and start timing. Place the thermometer in the silver can.

Step 4: At the 1-minute mark (1:00), read the temperature of the water in the silver can and record it in the data table. Quickly move the thermometer to the black can. Gently swirl the water in the can with the thermometer.

Step 5: At the 2-minute mark (2:00), read the temperature of the water in the black can and record it in the data table. Quickly move the thermometer to the white can. Gently swirl the water in the can with the thermometer.

Step 6: At the 3-minute mark (3:00), read the temperature of the water in the white can and record it in the data table. Quickly move the thermometer to the silver can. Gently swirl the water in the can with the thermometer.

Step 7: Repeat Steps 4–6 until the 21-minute mark temperature reading is made.

Data Table

Silver Can Temperatures (°C)	Black Can Temperatures (°C)	White Can Temperatures (°C)
Initial	Initial	Initial
T at 1:00	T at 2:00	T at 3:00
T at 4:00	T at 5:00	T at 6:00
T at 7:00	T at 8:00	T at 9:00
T at 10:00	T at 11:00	T at 12:00
T at 13:00	T at 14:00	T at 15:00
T at 16:00	T at 17:00	T at 18:00
T at 19:00	T at 20:00	T at 21:00

Step 8: Turn off the heat lamp. Plot your data for all three cans on a single temperature vs. time graph.

Step 9: Determine the change in temperature for the water in each can while the heat lamp was on.

a. Determine the temperature change in the silver can while the heat lamp was on. Subtract the 1-minute mark temperature reading from the 19-minute mark temperature reading.

Temperature change in silver can: _____°C

b. Determine the temperature change in the black can while the heat lamp was on. Subtract the 2-minute mark temperature reading from the 20-minute mark temperature reading.

Temperature change in black can: _____°C

c. Determine the temperature change in the white can while the heat lamp was on. Subtract the 3-minute mark temperature reading from the 21-minute mark temperature reading.

Temperature change in white can: _____°C

Summing Up

1. In which can did the water heat up at the fastest rate? In which can did the water heat up at the slowest rate? Did your observations match your predictions?

_____

2. Which would be a better choice if you were going to spend a long time outdoors on a hot, sunny day: a black T-shirt or a white T-shirt?

_____

3. What happens to the thermal radiation that falls on each of the cans: is it absorbed or reflected?

Silver: _____ Black: _____ White: _____

CONCEPTUAL PHYSICS	**Experiment**

Canned Heat: Cooling Down

Purpose
To compare the ability of different surfaces to radiate thermal energy

Apparatus
radiation cans: silver and black	paper towel
thermometer	access to hot water
graph paper	

Discussion
Does the color of a surface make a difference in how well it radiates thermal energy? If so, how? The answer to these questions could help you decide what color coffeepot will best keep its heat and what color might be used to radiate heat away from a computer chip. In this experiment, you will compare the thermal radiation ability of two surfaces: silver and black. You'll do this by filling two cans with these surfaces with hot water, then allowing them to cool down. You'll measure the temperature of the water in both cans while they're cooling down and see if there's a difference in the rate at which the temperatures decrease.

Pre-lab Questions
1. In which of the two cans do you think the water will cool down at the faster rate?

2. In which of the two cans do you think the water will cool down at the slower rate?

Procedure
Step 1: Carefully fill the cans with hot water and wipe up any spills. Quickly measure the initial temperature of the water in each of the cans and record it in the data table. Place the thermometer in the silver can.

Step 2: At the 1-minute mark, read the temperature of the water in the silver can and record it in the data table. Quickly move the thermometer to the black can. Gently swirl the water in the can with the thermometer.

Step 3: At the 2-minute mark, read the temperature of the water in the black can and record it in the data table. Quickly move the thermometer to the silver can. Gently swirl the water in the can with the thermometer.

Step 4: Repeat Steps 2–3 until the 20-minute mark temperature reading is made.

Data Table

Silver Can Temperatures (°C)	Black Can Temperatures (°C)
Initial	Initial
T at 1:00	T at 2:00
T at 3:00	T at 4:00
T at 5:00	T at 6:00
T at 7:00	T at 8:00
T at 9:00	T at 10:00
T at 11:00	T at 12:00
T at 13:00	T at 14:00
T at 15:00	T at 16:00
T at 17:00	T at 18:00
T at 19:00	T at 20:00

Step 5: Plot your data for both cans on a single temperature vs. time graph.

Step 6: Determine the change in temperature for the water in each can while it was allowed to cool.

a. Determine the temperature change in the silver can. Subtract the 1-minute mark temperature reading from the 19-minute mark temperature reading.

    Temperature change in silver can: _____°C

b. Determine the temperature change in the black can. Subtract the 2-minute mark temperature reading from the 20-minute mark temperature reading.

    Temperature change in black can: _____°C

Summing Up

1. In which can did the water cool down at the faster rate? In which can did the water cool down at the slower rate? Did your observations match your predictions?

---

2. What color should the surface of a coffeepot be in order to keep the hot water inside it hot for the longest time?

---

3. The central processing units (CPUs) in personal computers sometimes get very hot. To prevent damage and improve processing speed, they need to be kept cool. Often, a piece of metal is placed in contact with the CPU to draw some heat away. The metal then radiates its heat to the surroundings. To best remove heat from the CPU, should the metal's surface be colored black or left silvery?

CONCEPTUAL PHYSICS	**Demonstration**

I'm Melting, I'm Melting

Purpose
To observe the curious heat transfer abilities of different surfaces

Apparatus
defrosting tray
a second defrosting tray wrapped tightly in aluminum foil
white Styrofoam plate
black Styrofoam plate (or a white plate blackened completely using a felt-tip pen)
4 ice cubes of similar size
paper towel

Discussion
If you walk around inside your house with bare feet, you probably notice that a tile floor feels much colder than a carpeted floor or rug. It's hard to believe that they might actually have the same temperature. The tile feels colder because it is a better conductor than carpet. Heat is conducted from your warmer feet to the cooler floor faster when the floor is tile than when the floor is carpet. So your feet are cooled faster by tile than they are by carpet at the same cool temperature. In this activity, you will see which kinds of surfaces transfer heat most rapidly.

Procedure
Step 1: Set the defrosting tray, defrosting tray covered in aluminum foil, white Styrofoam plate, and blackened Styrofoam plate on your table.

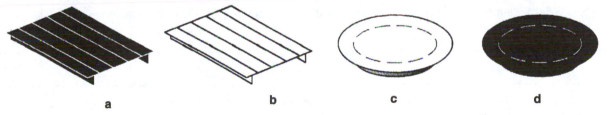

Figure 1. a. defrosting tray, **b.** defrosting tray covered in aluminum foil,
c. white Styrofoam plate, **d.** blackened Styrofoam plate

Which surfaces feel colder and which ones feel warmer? (Just touch a corner; don't transfer too much of your own body heat to any of the objects.)

_____

Step 2: In a moment, you'll set an ice cube on each of the surfaces. When you do, the cubes will begin melting. Before you set the ice cubes on the surfaces, make some predictions.

a.   Which ice cube will melt most quickly?

_____

b.   Which ice cube will melt most slowly?

_____

Step 3: Set the ice cubes out on their respective surfaces quickly. Observe the ice cubes for several minutes (preferably until the fastest-melting ice cube melts completely).

a. Which ice cube melted most quickly?

_____

b. Which ice cube will melt most slowly?

_____

Summing Up

1. How do your observations compare with your predictions?

_____

_____

2. Which way did heat flow in this activity? (From what to what?)

_____

3. Two of the surfaces were conductors and two of the surfaces were insulators. Which were which?

_____

_____

4. What advantage—in terms of heat transfer—did one defrosting tray have over the other?

_____

_____

5. What advantage—in terms of heat transfer—did one Styrofoam plate have over the other?

_____

_____

6. Which of the following conclusions are supported by your observations and which are not? Give evidence from this activity to justify your conclusion.

a. "Metals transfer heat faster than Styrofoam."

_____

b. "Black surfaces transfer heat faster than nonblack surfaces."

_____

CONCEPTUAL PHYSICS	**Demonstration**

Cooling by Boiling

Purpose
To see that water will boil when pressure is lowered

Apparatus
400-mL beaker
hot plate (or equivalent method for heating water above 60°C)
thermometer
vacuum pump with bell jar

Discussion
Whereas evaporation is a change of phase from liquid to gas at the surface of a liquid, boiling is a rapid change of phase at and below the surface of a liquid. The temperature at which water boils depends on atmospheric pressure. Have you ever noticed that water reaches its boiling point in a *shorter* time when camping up in the mountains? And have you noticed that at high altitude it takes *longer* to cook potatoes or other food in boiling water? That's because water boils at a lower temperature when the pressure of the atmosphere on its surface is reduced. Let's see!

Procedure
Step 1: Turn on the hot plate and warm 200 mL of water in a 400-mL beaker to a temperature above 60°C. Record the temperature. Then place the beaker under and within the bell jar of a vacuum pump. If a thermometer will fit underneath the bell jar, place a thermometer in the beaker. Turn on the pump. What happens to the water?

T = _____

_____

_____

Was the water *really* boiling?

_____

_____

Step 2: Stop the pump and remove the bell jar. What is the temperature of the water now?

T = _____

Step 3: As time permits, repeat the procedure, starting with other temperatures, such as 80°C, 40°C, and 20°C, recording the time it takes for boiling to begin.

Summing Up

1. In terms of energy transfer, what does it mean to say that boiling is a cooling process? What cools?

_____

_____

_____

_____

2. Name two ways to cause water to boil.

_____

_____

3. Boiling water on a hot stove remains at a constant 100°C temperature. How is this observation evidence that boiling is a cooling process?

_____

_____

_____

_____

4. In the photo opener to Chapter 17 in the textbook, and again in Figure 17.13, water is seen to freeze by rapid evaporation. How does this relate to this demonstration activity?

_____

_____

_____

CONCEPTUAL PHYSICS	Experiment

Warming by Freezing

Purpose
To measure the heat released when freezing occurs

Apparatus
heat-generating pouch (supersaturated sodium acetate pack)
hot plate
large pan or pot of boiling water

Discussion
To liquefy a solid or vaporize a liquid, it is necessary to add heat. In the reverse process, heat is released when a gas condenses or a liquid freezes. Thermal energy that accompanies these changes of state is called *latent heat of vaporization* (going from gas to liquid or liquid to gas) and *latent heat of fusion* (going from liquid to solid or solid to liquid).

Energy is absorbed when change of phase is in this direction

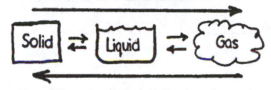

Energy is released when change of phase is in this direction

Water, which normally freezes at 0°C (32°F), can be found under certain conditions in a liquid state as low as –40°C (–40°F) or more. This *supercooled* water (liquid water below 0°C) often exists as tiny cloud droplets, common in clouds where snow or ice particles form. Freezing in clouds depends on the presence of *ice-forming nuclei*, most of which are active in the –10°C to –20°C range. Ice-forming nuclei comprise many different substances such as dust, bacteria, other ice particles, or silver iodide used to "seed" clouds during droughts. Silver iodide is active at temperatures as high as –4°C.

Cold clouds containing large amounts of supercooled water and relatively small amounts of ice particles can be dangerous to aircraft. The skin (outer layer) of the aircraft, well below freezing, provides an excellent surface on which supercooled water suddenly freezes. This is called aircraft icing, which can be quite severe under certain conditions.

The heat pack provides a dramatic example of a supercooled liquid. What you observe in the heat pack is actually the release of the latent heat of crystallization, which is analogous to the release of latent heat of vaporization or the latent heat of fusion. The freezing temperature of the sodium acetate solution inside the heat pack is about 55°C (130°F), yet it exists at room temperature. The heat pack can be cooled down to as low as –10°C before it finally freezes.

It only takes a quick click to activate the heat pouch. You will notice that the internal trigger button has two distinct sides. If you use your thumb and forefinger and squeeze quickly, you will not need to worry about which side is up.

After observing the crystallization of the sodium acetate and the heat released, you might want to try it again and measure the heat of crystallization. The package has a mass of about 146 grams. The packaging and the trigger mechanism have a mass of about 26 grams. Thus, the sodium acetate solution inside the package has a mass of about 120 grams.

Procedure

Place the heat pack in an insulated container (such as a larger Styrofoam cup) with 400 g of room-temperature water. Allow the water and heat pack to reach equilibrium. Measure and record its initial temperature. Activate the heat pack button. After 10 or 15 seconds, maximum temperature will be obtained. Return the heat pack to the container and measure the temperature of the water at regular intervals. How much heat is gained by the water?

Q = _____

Summing Up

1. How does the crystallization inside the heat pack relate to heating and air conditioning in a building? (*Hints:* Think about how steam heat and radiators operate or how the refrigerants in an air conditioner operate.)

_____

_____

_____

2. Speculate about how these processes relate to airplane safety?

_____

_____

_____

3. Think of and list some practical applications of the heat pouch.

_____

_____

_____

Wait a minute — before this we cooled by boiling, and now we warm by freezing? Must be some interesting physics going on here!

CONCEPTUAL PHYSICS	Demonstration

Slow-Motion Wobbler

Purpose
To observe and explore the oscillation of a tuning fork

Apparatus
low-frequency tuning fork (40–150 Hz forks work best for large amplitudes)
tuning fork mallet (or equivalent, optional)
bright strobe light with a widely variable frequency (inexpensive, party strobe lights **won't** do)
beaker (any size)
access to water

Discussion
The tines of a tuning fork oscillate at a very precise frequency. That's why musicians use them to tune instruments. In this activity, you will investigate their motion with a special illumination system—a **stroboscope.**

Procedure
Step 1: Strike a tuning fork with a mallet or the heel of your shoe (do **not** strike against the table or other hard object). Does it appear to vibrate? Try it again, this time dipping the tip of the tines just below the surface of water in a beaker. What do you observe?

_____

_____

Step 2: Now dim the room lights and strike a tuning fork while it is illuminated with a strobe light. For best effect, use the tuning fork with the longest tines available. Adjust the frequency of the strobe so that the tines of the tuning fork appear to be stationary. Then carefully adjust the strobe so that the tines slowly move back and forth.

a.   Describe your observations.

_____

_____

_____

_____

b. Which illustration below best depicts the motion of the tuning fork tines? Circle the correct sequence; cross out the incorrect sequence.

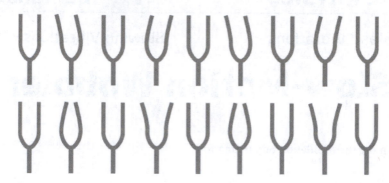

Step 3: Strike a tuning fork and observe how long it vibrates. Repeat, placing the handle against the tabletop or counter.

a. Is the sound louder or quieter?

_____

b. Does the *time* the fork vibrates increase or decrease? How does this make sense in terms of energy?

_____

_____

_____

Summing Up

1. What happens to the air next to the tines as they oscillate?

_____

_____

2. What would happen if you struck the tuning fork in outer space?

_____

_____

_____

| **CONCEPTUAL PHYSICS** | **Tech Lab** |

Water Waves in an Electric Sink

Purpose
To observe and control waves in a ripple tank simulation to learn the basics of wave mechanics

Apparatus
computer
PhET simulation: "Wave Interference" (available at http://phet.colorado.edu)

Discussion
The ripple tank is an effective (though cumbersome) classroom device used for demonstrating and exploring wave phenomena. A simple version is shown on page 511 in your textbook. More elaborate ones resemble a small glass table with raised edges. Water is poured onto the table and kept from spilling by the raised edges. Typically, a strong point light source is placed above the tank and shadows of ripples can be seen below the tank. A small ball attached to a motor bobs in and out of the water to make waves with consistent amplitude and wavelength.

Figure 1. A ripple tank

A variety of wave phenomena can be demonstrated using the ripple tank. This activity uses a ripple tank simulation, so you'll be able to investigate waves without the water.

Procedure

PART A: CRESTS AND TROUGHS

Step 1: When the simulation opens, you will see a faucet dripping water into a large sink. The drops create ripples in the water in the sink.

Step 2: Locate the "Rotate View" slider in the control panel on the right side of the window. Drag the slider to the right. Doing so rotates your view of the sink from a top view to a side view.

Step 3: Locate the "Pause" button at the bottom of the window. Try to pause the animation when the water under the faucet rises to its highest point (close to or touching the faucet itself).

Step 4: Locate the "Show Graph" button below the blue water of the ripple tank. Click it to activate the graph. Notice that the graph and the side view of the water match each other.

Step 5: Slide the "Rotate View" slider back to the left so that it shows the top view of the water.

a. In the spaces below, sketch the wave pattern as seen from the top and from the side.

Side View (Graph)	Top View

b. In both views (side view and top view), label a *crest* and a ***trough***.

c. In both views, label one wavelength.

PART B: AMPLITUDE

Step 1: Pause the animation. Locate the frequency slider below the faucet. Set the frequency to its maximum value by moving the slider all the way to the right. Restart the animation by clicking the on-screen "Play" button.

Step 2: Locate the amplitude slider. Slide it to various positions (to the left and right) and observe the effect this has on the simulation.

a. Does a change in amplitude result in a change in the size of the water drops? If so, how?

_____

_____

b. How are high-amplitude waves different from low-amplitude waves?

_____

_____

c. Review your sketches above (side view and top view) of the wave. Label the amplitude of the wave.

d. Which view—side or top—is better suited for labeling the amplitude? Explain?

_____

_____

e. What—if anything—happens to the amplitude of each wave as it gets farther away from the source?

_____

_____

PART C: FREQUENCY

Step 1: Pause the animation. Set the amplitude to its maximum value by moving the slider all the way to the right. Restart the animation by clicking the on-screen "Play" button.

Step 2: Move the frequency slider to various positions (to the left and right) and observe the effect this has on the simulation.

a. How are high-frequency waves different from low-frequency waves? (What *is* different?)

_____

_____

b. How are high-frequency waves the same as low-frequency waves? (What *isn't* different?)

_____

_____

c. Two students disagree about an observed difference between high-frequency waves and low-frequency waves. One says high-frequency waves are faster than low-frequency waves; the other claims both waves have the same speed. What do you think?

_____

_____

_____

_____

d. What is the relationship between the frequency (f) of the wave source (the dripping faucet) and the wavelength (λ) of the waves?

___ Direct proportionality: $\lambda \sim f$. The wavelength increases as the frequency increases.

___ Inverse proportionality: $\lambda \sim 1/f$. The wavelength increases as the frequency decreases.

___ No apparent relationship. The wavelength doesn't appear to be related to the frequency.

e. What—if anything—happens to the frequency of each wave as it gets farther away from the source?

_____

_____

Summing Up

1. Examine the illustrations below. Each represents a ripple tank wave. Some are side views; some are top views. Describe the amplitude of the wave and the frequency of its source by using the terms "high" or "low." Please examine all the patterns before recording your descriptions. (*Hint:* Waves a–d are all different from one another.)

a. _____amplitude

   _____ frequency

Water Waves in an Electric Sink

b. _____amplitude

_____ frequency

c. _____amplitude

_____ frequency

d. _____amplitude

_____ frequency

2. What single aspect of a wave does its amplitude best represent?

____ speed   ____ wavelength   ____ frequency   ____ energy   ____ period

3. a. Which control on a music player or television set allows you to increase or decrease the amplitude of the sound waves that come out of it?

_____

Recall what happens to the amplitude of a wave as the wave gets farther from the source. Imagine a portable music player playing music in a large, open field. At some distance from the player, the amplitude of the sound waves diminishes to zero, and the sound cannot be heard. Consider the 3-dimensional space in which the sound can be heard.

b. How might you increase that space, and what is 3-dimensional space called (in geometry)?

_____

_____

c. Ripple tanks are used to observe 2-dimensional waves. What should be the name of the amplitude control for 2-dimensional waves?

_____

_____

CONCEPTUAL PHYSICS	Tech Lab

High Quiet Low Loud

Purpose

To use sound generation software to manipulate the amplitude and wavelength of a sound wave, and observe the connections between the plot of a sound wave and the characteristics of the sound

Apparatus

computer
sound generation software (Pasco's DataStudio with Waveport or equivalent)
headphones (optional: in-line volume control)
signal splitters (if two or more lab partners are connected to a single computer)

Discussion

Sound is a mechanical wave. It is a longitudinal wave that travels through materials. Sound is generated by a vibrating source and is received by objects that can also vibrate, such as the tympanic membrane (the eardrum). Like all mechanical waves, sound waves have amplitude and wavelength. Your sense of hearing is able to distinguish sound waves by their tone (or pitch) and volume (or loudness). Wavelength and amplitude are related to volume and tone. In this activity, you will find the specific connections.

Procedure

Step 1: If it's not already running, start the computer.

Step 2: While the computer is starting, connect the headphones to the signal splitters as needed. Every member of the lab group should have his or her own pair of headphones.

Step 3: Make sure the computer's sound is on and turned all the way up.

Step 4: Start the sound generation software (e.g., DataStudio with Waveport).

Step 5: Activate the sound-generating component of the software (e.g., SoundCreator). If the software has a separate analysis component (e.g., SoundAnalyzer), close it for now. If you have difficulty, ask your instructor for assistance.

Step 6: Start the sound by, for example, clicking the on-screen image of the speaker. Examine the initial sound as graphically displayed by the software.

Notice that the *longitudinal* sound wave is represented here as a *transverse* wave. The displayed wave actually represents the electrical signal used to drive the speaker. Current in one direction pushes the speaker cone out; current the other way pulls the speaker cone in. The sinusoidal current pushes and pulls the speaker cone many times each second to create sound waves in the air.

Any changes made to the electrical signal will change the resulting sound.

Step 7: Make the wave *smaller or taller* by altering the amplitude of the wave. There should be an on-screen control (such as a hand icon) that you can move by clicking and dragging.

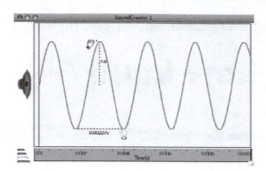 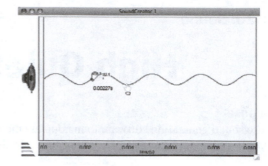

Figure 1. Making the wave taller or smaller by manipulating the amplitude.

What effect does changing the signal's amplitude have on the sound that you hear? Be specific: What does increasing the amplitude do, and what does decreasing the amplitude do?

_____

_____

_____

Step 8: Make the wave *longer or shorter* by altering the wavelength of the wave. There should be an on-screen control (such as a hand icon) that you can move by clicking and dragging.

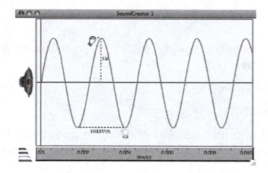 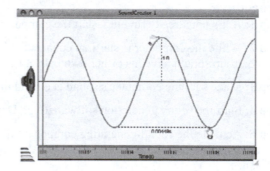

Figure 2. Making the wave longer or shorter by manipulating the wavelength.

What effect does changing the signal's wavelength have on the sound that you hear? Be specific: What does increasing the wavelength do, and what does decreasing the wavelength do?

_____

_____

_____

Summing Up

1. Consider the first wave to be a wave with a medium volume (loudness) and average tone (pitch).

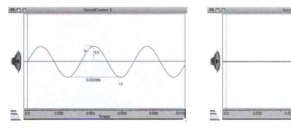

a.

b.

Draw:
   a. the signal for a louder-sounding wave with roughly the same pitch.
   b. the signal for a quieter-sounding (but not silent) wave with roughly the same pitch.

2. Consider the first wave to be a wave with a medium volume (loudness) and average tone (pitch).

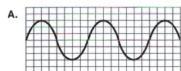
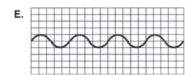
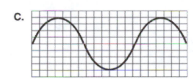
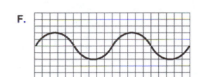

a.

b.

Draw:
   a. the signal for a higher-pitched wave with roughly the same volume.
   b. the signal for a lower-pitched wave with roughly the same volume.

3. The waves illustrated below represent signals for speakers.

A.
B.
C.
D.
E.
F.

a. Rank them from loudest to quietest.

_____

b. Rank them from highest to lowest.

_____

High Quiet Low Loud

CONCEPTUAL PHYSICS **Experiment**

Fork It Over

Purpose
To use simple materials to determine the speed of sound in air

Apparatus
resonance tube apparatus: 250-mL or larger graduated cylinder, matching length of PVC tube
tuning fork with a known frequency (approximately 300 Hz–500 Hz)   meterstick
tuning fork mallet or wedge activator   access to water
access to thermometer   paper towel

Going Further
tuning fork with an "unknown" frequency value

Discussion
The standing wave is an example of resonance. Standing waves can arise in stretched cords, such as guitar, violin, or piano strings. They can also arise in air columns, such as those in organ pipes. The simplest form of a standing wave—the one that arises at the lowest resonant frequency—is called the ***fundamental.***

When a fundamental arises in a medium constrained by two fixed ends, the resonant frequency corresponds to a wavelength twice as long as the distance between the ends. When a fundamental arises in a medium with two free ends, the resonant frequency also corresponds to a wavelength twice as long as the distance between the ends.

But when a fundamental arises in a medium constrained by one fixed end and one free end, the resonant frequency corresponds to a wavelength four times as long as the distance between the ends.

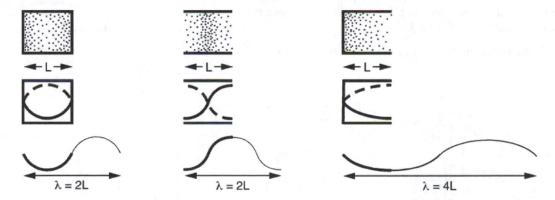

Figure 1.a. When the resonant length has two fixed ends or two free ends, the resonant wavelength is 2L.

Figure 1.b. When the resonant length has one fixed end and one free end, the resonant wavelength is 4L.

In this experiment, a tuning fork will be used to set air into oscillation. A tube partially submerged in water will allow you to control the length of a column of air. When the tuning fork oscillates the air in the column, and the column is adjusted to the correct length, resonance will occur. Evidence of resonance in the column of air will result in amplification of the tone produced by the tuning fork.

The resonant column of air will have a fixed end at the water and a free end in the opening at the top.

Pre-lab Questions

1. a. Does the resonating column of air in this experiment have two fixed ends, two free ends, or one fixed end and one free end?

_____

    b. How long is the resonant **wavelength** compared with the **length** of the resonant air column?

_____

2. If a 400-Hz tuning fork produced resonance in an air column 22 cm high,
    a. what would be the resonant wavelength (in meters)?

_____

    b. what would be the speed of sound in air for such an observation? Use the wave equation, $v = f\lambda$, and the knowledge of the frequency of the tuning fork and resonant wavelength.

_____

_____

3. Ask your instructor to approve your Pre-lab work before moving on to the Procedure.

Procedure

Step 1: Record the frequency of the tuning fork.

f = _____ Hz

Step 2: With the PVC tube inserted, fill the graduated cylinder with water. Leave the top 2 to 3 centimeters empty to avoid spilling.

Step 3: Using the mallet or wedge, strike the tuning fork and hold it near the open end of the PVC tube as shown in Figure 2.

Step 4: Carefully raise the PVC tube up out of the water while listening to the sound. Keep the tuning fork near the open top of the tube. Locate the resonant point, the point at which the sound in the tube is loudest. If you overshoot the resonant point, move the tube back downward. (See Figure 3.)

Reactivate the tuning fork as needed during the process.

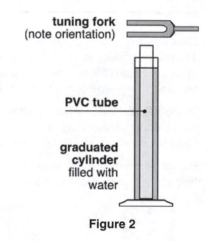

tuning fork
(note orientation)

PVC tube

graduated
cylinder
filled with
water

Figure 2

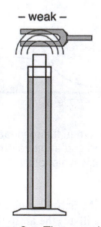

Figure 3.a. The sound is faint when the air column is too short.

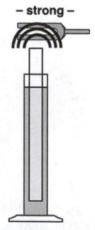

Figure 3.b. The sound is loud when the air column is just right!

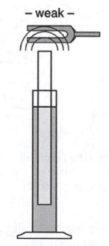

Figure 3.c. The sound is faint when the air column is too long.

Step 5: When you are confident that you have identified the lowest resonant point, measure and record the distance from the water level to the open end of the PVC tube. This is the length L of the resonant air column. See Figure 4.

Step 6: Record the air temperature in the classroom (in degrees Celsius).

T = _____ °C

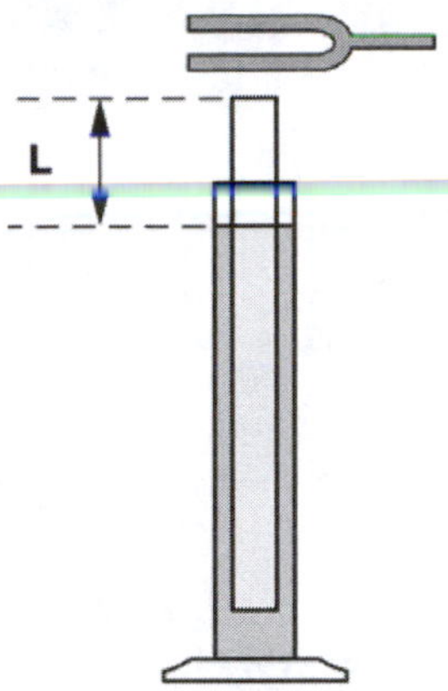

Summing Up

1. In terms of the length L of the resonant air column, what is the resonant wavelength λ at resonance? (Write an equation, not a numerical value.)

_____

Figure 4. The length L of the resonant air column.

2. Determine the speed of sound in air based on the resonant wavelength. Show your work.

_____

_____

_____

3. The speed of sound in air is given to be 331 m/s at 0 °C and rises 0.6 m/s for every Celsius degree. Determine the speed of sound in air based on temperature. Show your work!

_____

_____

4. Determine the percent *error*. If you are not sure how to calculate percent error, refer to Appendix C on percent error and percent difference at the end of this manual.

_____

_____

Going Further

Ask your instructor for a tuning fork with an unknown frequency.

1. Use the techniques of this experiment to determine the frequency f_{LAB} of the tuning fork. Record your data and calculations on a separate sheet. Record the frequency value you find.

f_{LAB} = _____ Hz

2. Ask your instructor for the accepted value of the frequency f_{ACC} of the tuning fork. Record the value below.

f_{ACC} = _____ Hz

3. Determine the percent error. Show your work.

Fork It Over

Science is built up with facts, as a house is with stones.
But a collection of facts is no more a science than a heap of stones is a house.

Jules Henri Poincaré

CONCEPTUAL PHYSICS	Demonstration

Sound Off

Purpose
To observe and interpret a dramatic effect of the interference of sound

Apparatus
Stereo radio, tape, or CD player with two moveable speakers, one of which has a DPDT (double pole double throw) switch *or* a means of reversing polarity (such as switching red and black wires)

Discussion
Interference is a behavior common to all waves. With water waves we see it in regions of calm where overlapping crests and troughs coincide. We see the effects of interference in the colors of soap bubbles and other thin films where reflection from nearby surfaces puts crests coinciding with troughs. In this activity you'll carry out the dramatic demonstration of sound interference as shown in Figure 20.19 in your textbook. Amazing!

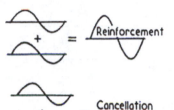

Procedure
Play the stereo player with both speakers in phase (with the plus and minus connections to each speaker the same). Play it in mono mode so the signals of each speaker are identical. Note the fullness of the sound. Now reverse the polarity of one of the speakers (either by physically interchanging the wires or by means of the switch provided). Note the sound is different—it lacks fullness. Some of the waves from one speaker are arriving at your ear out of phase with waves from the other speaker.

Now place the speakers facing each other at arm's length. The long waves are interfering destructively, detracting from the fullness of the sound. Gradually bring the speakers closer to each other. What happens to the volume and fullness of the sound heard? Bring them face-to-face against each. What happens to the volume now?

Summing Up
1. What happens to the volume of sound when the face-to-face speakers are switched so both are in phase?

    _____

2. Why is the volume so diminished when the out-of-phase speakers are brought together face-to-face? And why is the remaining sound so *tinny?*

    _____

3. What practical applications can you think of for canceling sound?

    _____

Going Further

1. Research "Active Sound Control," a feature used in some models of luxury automobiles. Describe the connections to this demonstration.

_____

_____

_____

_____

_____

2. Research "noise cancelling headphones." Describe the connections to this demonstration.

_____

_____

_____

_____

CONCEPTUAL PHYSICS

Tech Lab

Wah-Wahs and Touch-Tones

Purpose
To use sound generation software to produce and observe the interference of multiple sound waves

Apparatus
computer
sound generation software (Pasco's DataStudio with Waveport or equivalent)
headphones (optional: in-line volume control)
signal splitters (if two or more lab partners are connected to a single computer)

Going Further
access to the Internet

Discussion
When two sound waves interfere with one another, the resulting sound pattern is referred to as **beats.** The beat pattern is determined by the waves that produce it. Piano tuners use knowledge of beats in their practice. A tuning fork with a key's frequency is struck and the note is played. If beats are heard, an adjustment needs to be made. Telephones send tones when keys are pressed. The tones are combinations of two notes played at the same time. Beats are all around us!

Procedure

SETUP
Step 1: If it's not already running, start the computer.

Step 2: While the computer is starting, connect the headphones to the signal splitters as needed. Every member of the lab group should have his or her own pair of headphones.

Step 3: Make sure the computer's sound is on and turned all the way up.

Step 4: Start the sound generation software (e.g., DataStudio with Waveport).

Step 5: Activate the sound-generating component of the software (e.g., SoundCreator). If the software has a separate analysis component (e.g., SoundAnalyzer), close it for now. If you have difficulty, ask your instructor for assistance.

Step 6: Start the sound by, for example, clicking the on-screen image of the speaker. Consider the image below to represent the initial sound as displayed by the software. Any changes made to the electrical signal will change the resulting sound.

Step 7: Double-click in the SoundCreator window to activate the SoundCreator "Settings" window.
a. The "Appearance Pane" should open by default. Type a "2" into the "Simultaneous Tones" box ("1" is the default value).
b. Click the "Tones Display" tab. Click the checkbox to activate the "Phase Tool." Click the checkbox to "Show Frequency Buttons." Click "OK" to close the "Settings" window.

PART A: PHASE
Step 1: Click the on-screen speaker icon in the lower window to activate the tone.

The tone consists of two input waves. The two input waves are depicted in the upper window. One is red; one is green. Colored buttons to the left of the waveforms allow the user to select an input wave.

The result of the two waves is depicted in the lower window. The two colored dots near the speaker icon show the input waves. Arrows from the dots to the speaker indicate which input waves are active.

Step 2: In the lower window, click the green dot to deactivate the green input wave.

What effect—if any—does removing one input wave have on the resulting sound?

_____

_____

Step 3: Use the phase tool (on-screen hand icon farthest to the left in the upper window) to change the phase of the red input wave. Adjust the phase forward and back while listening to the tone.

What effect—if any—does changing the phase of a wave have on the resulting sound?

_____

_____

PART B: INTERFERENCE

Step 1: Return the phase of the red wave back to its original state. Reactivate the green wave by clicking the green dot near the speaker. The lower window and the resulting sound should reflect that there are now two input waves.

Step 2: Now adjust the phase of the green wave in the upper window and observe the resulting sound.

Describe, using words and pictures, the result when the two waves are completely out of phase. Figure 1.a. shows the result when the waves are completely in phase. Figure 1.b. shows the two input waves completely out of phase (but does not show the result).

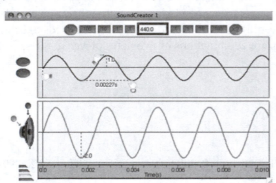
Figure 1.a.

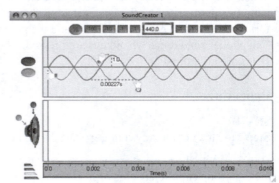
Figure 1.b.

PART C: BEATS

Step 1: Return the waves to their original state (in phase).

If you ever have trouble resetting the input waves to their original state, close the SoundCreator window and create a new one. Then adjust the settings as described in Steps 1–7.

Step 2: Drag the on-screen wavelength tool (hand icon connected to the trough) to adjust the frequency of one of the input waves.

What is the effect when two input waves have slightly different frequencies?

_____

_____

Step 3: Using the frequency control buttons, adjust one input wave to have a frequency of 440.0 Hz and the other to have a frequency of 441.0 Hz.

Step 4: Listen for the beat frequency. The beat frequency is equal to the difference between the frequencies of the input waves. In this case, the beat frequency is 441.0 Hz – 440.0 Hz = 1.0 Hz. You should hear a 1.0-Hz "wah-wah" pattern in the resulting sound.

How else can you generate a 1.0-Hz beat frequency without changing the frequency of the 440.0-Hz input wave?

_____

_____

Step 5: Experiment with other combinations of input wave frequencies to create beats. Low frequencies produce interesting results.

Summing Up

1. How do the terms *constructive interference* and *destructive interference* relate to the relative phase of the two input waves?

When two waves are in phase, _____ interference occurs.

When two waves are out of phase, _____ interference occurs.

2. What is the beat frequency when the adjacent piano notes C (262 Hz) and C-sharp (277 Hz) are played simultaneously? Show your work.

_____

_____

Going Further: Touch-Tones

Dual-tone multifrequency (DTFM) phone tones use two frequencies for each key. For example, when the "1" key is pressed, the tone that plays is the combination of 697 Hz and 1209 Hz. The resulting beat frequency is 1209 Hz – 697 Hz = 512 Hz.

1. Access the Internet to research the frequency pairs for all the other keys on a telephone keypad.

1 = 697 + 1209	2 = _____ + _____	3 = _____ + _____
4 = _____ + _____	5 = _____ + _____	6 = _____ + _____
7 = _____ + _____	8 = _____ + _____	9 = _____ + _____
* = _____ + _____	0 = _____ + _____	# = _____ + _____

2. Determine the beat frequency for each key.

1 = 512 Hz	2 = _____	3 = _____
4 = _____	5 = _____	6 = _____
7 = _____	8 = _____	9 = _____
* = _____	0 = _____	# = _____

He who proves things by experience increases his knowledge;
he who believes blindly increases his errors.

Chinese proverb

CONCEPTUAL PHYSICS	Activity

A Force to Be Reckoned

Purpose
To explore the nature of a new force and determine whether or not it is distinct from other known forces

Apparatus
a pith ball suspended from a support rod assembly (as shown below)

vinyl (opaque plastic) strips or tubes   acetate (transparent plastic) strips or tubes
wool (heavy, coarse) cloth squares (~ 4" × 4")   silk (lightweight, smooth) cloth squares
brick (or a heavy book)   bar magnet

Going Further
electrophorus   second electrophorus or pie tin

Discussion
When we see an object accelerate, we know there must be an unbalanced, external force acting on it. We have seen mechanical forces at work in previous lessons. Weight, friction, tension, normal force, lift, and drag are now familiar to us. In this activity, we will see a force that may seem new to us. Before we treat it as a different force, we need to make sure it's not a force we're already familiar with.

Procedure
Step 1: Generate an attractive force. Rub the vinyl with the wool. Hold the vinyl near the pith ball to see if the strip attracts the pith ball to it as shown. If it doesn't, try rubbing the vinyl more vigorously, or try using the acetate rubbed with silk. If you continue to have difficulty, ask your instructor for assistance.

Once you have observed the plastic strip (vinyl or acetate) attract the pith ball, set the plastic strip aside and discharge the pith ball by gently touching it.

Step 2: Can you reasonably conclude that the attraction you observed was not due to gravitational force? Hold the brick or heavy book close to the pith ball. What—if anything—does the brick do to the pith ball? What does this tell you about the notion that the attraction between the plastic strip and the pith ball is *gravitational?*

_____

_____

_____

_____

Step 3: Generate a repulsive force. Using the plastic strip (rubbed with the cloth as was done in Step 1), try to repel the pith ball as shown. Does the observation of repulsion support or contradict the suggestion that the attraction between the plastic strip and the pith ball is gravitational? Why?

_____

_____

_____

_____

Once you have observed the plastic strip (vinyl or acetate) attract the pith ball, set the plastic aside and discharge the pith ball by touching it.

Consider the suggestion that the attraction and repulsion you observed are simply the result of *magnetic* force. After all, magnets can both attract *and* repel.

Step 4: Hold the bar magnet close to the pith ball. What does the magnet do to the pith ball? What does this tell you about the suggestion that the attraction and repulsion between the plastic strip and the pith ball are magnetic?

_____

_____

Summing Up

The force between the plastic strip and the pith ball is different from gravitational force and different from magnetic force. It is called *electrostatic* force, and is a force between any two objects that are electrically charged.

1. When the vinyl is rubbed with wool, the vinyl gets a *negative* charge. What kind of charge is on the pith ball when the charged vinyl repels it?

_____

2. Does electrostatic force get stronger or weaker with distance? Does the interaction become stronger when charged objects get closer together or when they get farther apart?

_____

_____

3. A student doing this activity argues that the attractive force observed in Step 1 cannot be a gravitational force because gravitational force is always vertical (downward) and cannot be horizontal. Is this a good argument? Explain.

_____

_____

_____

_____

Going Further

PART A: ELECTROPHORUS CHARGE

Step 1: Touch the pith ball to neutralize it.

Step 2: Charge the electrophorus. See Figure 1 below

 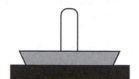  

Figure 1. Charging the Electrophorus

a. Rub the electro- phorus **base** or a plastic chair seat with the appropriate cloth (silk on acylic; wool on vinyl or Styrofoam).

b. Set the electro- phorus **plate** on the base or chair seat. Avoid metal rivets in the seats.

c. Touch the top of the metal plate. You may feel a slight shock; you may feel nothing. **Do not skip this step!**

d. Pick up the plate holding only the plastic handle. If the plate discharges, repeat the previous steps.

Step 3: With the electrophorus plate horizontal, move the plate toward the suspended pith ball. The pith ball should be attracted to the electrophorus plate, touch it, then be repelled from it. The pith ball is now charged with the same type of charge the electrophorus held (*like* charges repel).

Step 4: Does wool-rubbed vinyl attract or repel the charged pith ball? (It **should** do one or the other.) (Note: If there is attraction and the pith ball **touches** the vinyl, you'll need to repeat Steps 1–3 before proceeding to Step 5.)

_____

Step 5: Does silk-rubbed acetate attract or repel the electrophorus-charged pith ball?

_____

Step 6: What does this tell you about wool-rubbed vinyl and silk-rubbed acetate?

_____

_____

_____

A Force to Be Reckoned

PART B: PITH BALL PING-PONG

Step 1: Touch the pith ball to neutralize it.

Step 2: Charge the electrophorus and hold it properly by its handle.

Step 3: With the charged electrophorus in one hand and an uncharged electrophorus (or pie tin) in the other, hold the two so that they are parallel and vertical, as shown below. Use them to slowly sandwich the charged pith ball. If all goes well, a nice "game" of pith ball Ping-Pong should ensue.

If you are using a second electrophorus for your uncharged object, hold it by the rim (instead of by the handle) for best results. The charged electrophorus *must* be held by the handle.

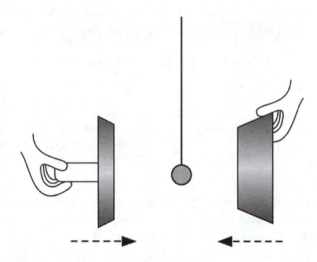

CONCEPTUAL PHYSICS	Experiment

Chapter 22: Electrostatics   Conduction, Insulation, Conductors, and Insulators

Electroscopia

Purpose

To use an electroscope to investigate electric charge polarity, electrostatic induction, and the difference between conductors and insulators

Apparatus

electroscope (can-form, *or* flask-form, *or* gold leaf)   electrophorus
2 acetate strips (transparent plastic)   2 vinyl strips (opaque plastic)
silk cloth square (thin and soft)   wool cloth square (thick and coarse)
small metal tube or rod   hair dryer (optional)

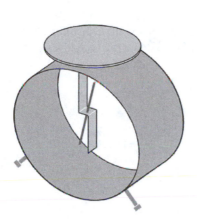

Can-form Electroscope

Flask-form Electroscope

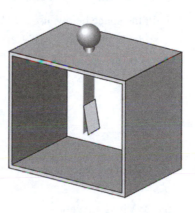

Gold Leaf Electroscope

Discussion

A microscope allows you to see things that are very small, and a telescope allows you to see things that are very far away. An electroscope allows you to detect the presence of excess electric charge. Your lab likely has *one* of the three types of electroscopes shown above. All three perform the same function. The nature of electric charge, charge transfer, and conductors and insulators can be investigated through the use of the electroscope.

Procedure

PART A: POLARITY

Step 1: Check to see that the electroscope and the charging materials are functioning correctly.

a. When the electroscope is neutral and no charged objects are nearby, its pointer or foil leaf(s) should be "undeflected." See the simplified illustrations in Figure 1.

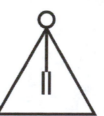

Figure 1. Neutral Electroscope
Can-form, flask-form, and gold leaf

Note: You'll have only one type of electroscope. When you make your own illustrations later, use the diagram that corresponds to your electroscope type.

b. When a vinyl strip is vigorously rubbed with wool cloth and brought nearby, the pointer or leaf(s) should deflect. See the simplified illustrations in Figure 2. The deflected electroscope indicator shows the presence of excess electric charge.

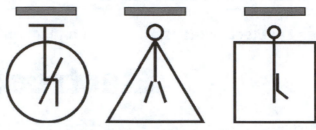

Figure 2. Electroscope with Nearby Charge
Can-form, flask-form, and gold leaf

c. When an acetate strip is rubbed vigorously with silk cloth and brought nearby, the electroscope indicator should again deflect.

Step 2: Vinyl + Vinyl

a. Rub a strip of vinyl with wool and hold it near the top of the electroscope. Keep it there.

b. While the first vinyl strip remains near the top of the electroscope, rub a second vinyl strip with wool and hold it near the first strip, close to the top of the electroscope.

c. In the space below, use words and illustrations to record your observations. (Show the type of electroscope you are using and sketch two illustrations—before and after—to show how adding the second strip affects the electroscope.)

Step 3: Vinyl + Acetate

a. Rub a strip of vinyl with wool and hold it near the top of the electroscope. Keep it there.

b. While the vinyl strip remains near the top of the electroscope, rub an acetate strip with silk and hold it near the vinyl strip, close to the top of the electroscope.

c. In the space below, use words and illustrations to record your observations. (Sketch before and after illustrations to show how adding the second strip affects the electroscope.)

Step 4: Acetate + Acetate

a. Rub a strip of acetate with silk and hold it near the top of the electroscope. Keep it there.

b. While the first acetate strip remains near the top of the electroscope, rub a second acetate strip with silk and hold it near the first strip, close to the top of the electroscope.

c. In the space below, use words and illustrations to record your observations. (Show the type of electroscope you are using and sketch before and after illustrations.)

Step 5: Acetate + Vinyl

a. Rub a strip of acetate with silk and hold it near the top of the electroscope. Keep it there.

b. While the acetate strip remains near the top of the electroscope, rub a vinyl strip with wool and hold it near the acetate strip, close to the top of the electroscope.

c. In the space below, use words and illustrations to record your observations. (Show the type of electroscope you are using and sketch before and after illustrations.)

Step 6: Interpret these observations.

a. When one vinyl charge is added to another vinyl charge, do they tend to enhance one another or cancel one another? Answer and describe your electroscope-based evidence.

_____

_____

b. When one acetate charge is added to another acetate charge, do they tend to enhance one another or cancel one another? Answer and describe your electroscope-based evidence.

_____

_____

c. When vinyl and acetate charges are added to one another, do they tend to enhance one another or cancel one another? Answer and describe your electroscope-based evidence.

_____

_____

> **Scientists refer to two types of charge. For historical reasons, the charge on silk-rubbed acetate is called _positive_, and the charge on wool-rubbed vinyl is called _negative_.**

Electroscopia

PART B: DETERMINING THE SIGN OF AN UNKNOWN CHARGE

Step 1: Based on what you learned in Part A, describe a method for determining the type (sign) of charge of a charged object. You know the charge of wool-rubbed vinyl, and you know the sign of silk-rubbed acetate. But how can you determine the charge of any object using what you've learned so far?

_____

_____

_____

_____

Step 2: Charge the electrophorus. See Figure 3. *Pay very close attention to the details described in each step, and follow the sequence exactly as described.*

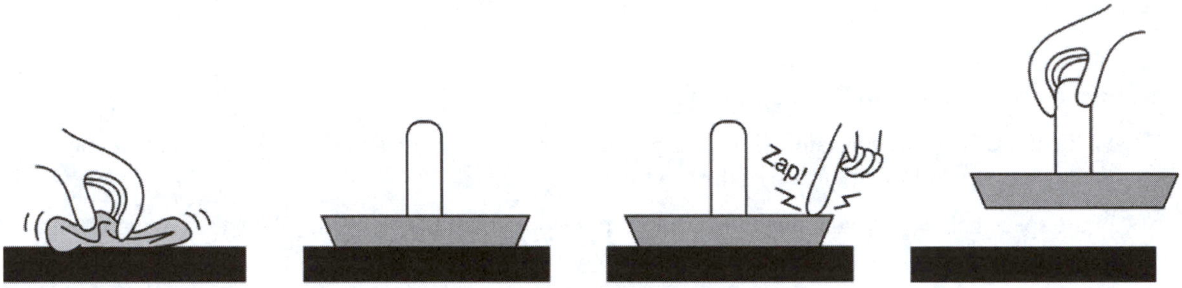

Figure 3. Charging the Electrophorus

a. Rub the electro-phorus **base** or a plastic chair seat with the appropriate cloth (silk on acylic; wool on vinyl or Styrofoam).

b. Set the electro-phorus **plate** on the base or chair seat. Avoid metal rivets in the seats.

c. Touch the top of the metal plate. You may feel a slight shock; you may feel nothing. **Do not skip this step!**

d. Pick up the plate holding only the plastic handle. If the plate discharges, repeat the previous steps.

Step 3: Carry out the procedure you described in Step 1 above to determine the charge on your electrophorus. What is the charge on the electrophorus and what is your evidence?

_____

_____

_____

Step 4: Verify that your finding is correct. If you used an acrylic (transparent plastic) base, then your electrophorus plate carried a negative charge. If you used a vinyl (opaque plastic) or Styrofoam base, then your electrophorus plate carried a positive charge.

Step 5: Ask your instructor to bring an "unknown charged *object*" near your electroscope for investigation.
a.   Describe the object.

_____

_____

b.   What is the sign of the charge on the object and what is your evidence?

_____

_____

_____

Step 6: Describe a method for determining the type (sign) of charge of a charged electroscope. In other words, if your electroscope indicator were deflected with no charged objects nearby, how could you determine the sign of the excess charge on the electroscope?

_____

_____

_____

_____

Step 7: Neutralize your electroscope by touching the top plate or metal ball. Ask your instructor to transfer an "unknown *charge*" onto your electroscope.
a. Describe the method by which your instructor charged your electroscope.

_____

_____

b. What was the sign of the charge transferred to your electroscope and what is your evidence?

_____

_____

_____

PART C: ELECTROSTATIC INDUCTION
Step 1: Neutralize your electroscope so that the indicator is undeflected.

Step 2: Recall and record the sign of the charge on wool-rubbed vinyl: _____ .

Step 3: Rub a strip of vinyl with wool and bring it near to the top plate or ball of the electroscope. Observe the corresponding deflection of the indicator.

Step 4: While keeping the vinyl in place near the electroscope, *briefly* touch the top plate or ball of the electroscope. Observe that the indicator reverts to its undeflected state.

Step 5: Now move the vinyl away from the electroscope. What do you observe?

_____

_____

Step 6: Illustrate the entire sequence in the space below.

| a. The electroscope is neutral; the indicator is undeflected. | b. Wool-rubbed vinyl is brought nearby. | c. With the vinyl in place, the top plate or ball of the electroscope is briefly touched. | d. The vinyl is removed. |

Step 7: Using your procedure from Step 6 of Part B: Determining the Sign of an Unknown Charge, determine the sign of the charge left on the electroscope.

a. What is the sign of the charge left on the electroscope?   ___positive   ___negative

b. How does this compare to the sign of the charge on the vinyl?   ___the same   ___opposite

> **Charging an object without direct contact is called *induction*.**

PART D: CONDUCTORS AND INSULATORS

There is something different between conductors and insulators. Perhaps several things. Plastics, such as vinyl and acetate, are insulators. Metals, such as iron and aluminum, are conductors.

Step 1: Select the best descriptions of conductors and insulators, based on what you've observed so far.

___Conductors can be charged.   ___Insulators can be charged.

___Conductors are always charged.   ___Insulators are always charged.

___Conductors are never charged.   ___Insulators are never charged.

Step 2: Neutralize the electroscope by touching the top metal plate or ball. Charge a strip of plastic (vinyl or acetate—whichever works better for you). Place the charged plastic strip in direct contact with the top metal plate or ball of the electroscope. Then move the strip away from the electroscope and set it aside. In the space below, describe what happened using words and illustrations.

Step 3: Neutralize the electroscope by touching the top metal plate or ball. Charge the electrophorus (using the acrylic base or chair seat—whichever works better for you). Place the charged metal plate in direct contact with the top metal plate or ball of the electroscope. Then move the electrophorus plate away from the electroscope and set it aside. In the space below, describe what happened using words and illustrations.

Step 4: Describe the important difference in behavior between the charged insulator (plastic strip) and the charged conductor (metal plate) in the previous steps.

_____

_____

_____

> **Charging an object by direct contact is called *conduction*.**

Summing Up

1. An object is brought near the top metal plate or ball of a neutral electroscope. The indicator deflects. From this, we may conclude that (select all that apply)
   ____the object carries a positive charge.   ____the object carries a negative charge.
   ____the object carries a charge, but the sign of the charge is unknown.
   ____the object is a conductor.   ____the object is an insulator.
   ____the object may be either a conductor **or** an insulator.

2. An object is brought near the top metal plate or ball of a neutral electroscope. The indicator deflects. The object touches the top of the electroscope and is then removed. The indicator on the electroscope remains deflected. From this, we may conclude that (select all that apply)
   ____the object carries a positive charge.   ____the object carries a negative charge.
   ____the object carries a charge, but the sign of the charge is unknown.
   ____the object is a conductor.   ____the object is an insulator.
   ____the object may be either a conductor **or** an insulator.

3. A strip of charged acetate is held near a neutral electroscope, causing the indicator to deflect. An object with an unknown charge approaches the top of the electroscope, and the indicator's deflection ***increases***. From this, we may conclude that (select all that apply)
   ____the object carries a positive charge.   ____the object carries a negative charge.
   ____the object carries a charge, but the sign of the charge is unknown.
   ____the object is a conductor.   ____the object is an insulator.
   ____the object may be either a conductor **or** an insulator.

4. An object is brought near the top metal plate or ball of a neutral electroscope. The indicator deflects. The metal plate or ball at the top of the electroscope is briefly touched. The object is then removed. The indicator on the electroscope remains deflected. From this, we may conclude that (select all that apply)
   ____the object carries a positive charge.   ____the object carries a negative charge.
   ____the object carries a charge, but the sign of the charge is unknown.
   ____the object is a conductor.   ____the object is an insulator.
   ____the object may be either a conductor **or** an insulator.

5. A strip of charged vinyl is held near a neutral electroscope, causing the indicator to deflect. An object with an unknown charge approaches the top of the electroscope, and the indicator's deflection ***decreases***. From this, we may conclude that (select all that apply)
   ____the object carries a positive charge.   ____the object carries a negative charge.
   ____the object carries a charge, but the sign of the charge is unknown.
   ____the object is a conductor.   ____the object is an insulator.
   ____the object may be either a conductor **or** an insulator.

6. An object is brought near the top metal plate or ball of a neutral electroscope. The indicator deflects. The object is then removed, touched, and returned. The indicator on the electroscope again deflects. From this, we may conclude that (select all that apply)
   ____the object carries a positive charge.   ____the object carries a negative charge.
   ____the object carries a charge, but the sign of the charge is unknown.
   ____the object is a conductor.   ____the object is an insulator.
   ____the object may be either a conductor **or** an insulator.

7. Based on your observations in this experiment, what is the difference in the behavior of conductors and insulators?

_____

_____

_____

8. **Charging the Electrophorus**
The electrophorus is charged by induction. Show the charges in the electrophorus plate in Figure 4 to fully illustrate the movement of charge during this process.

Figure 4. Charging the Electrophorus by Induction

a. When the electrophorus plate is placed on a negatively charged base, charge within the plate separates. The plate becomes polarized. (Positive charge is attracted to the base; negative charge is repelled from it.)

b. When the electrophorus is touched, negative charge escapes to you and on to the Earth; positive charge stays with the electrophorus.

c. The electrophorus is left with a charge opposite to that of the charged base.

9. Suppose that charging an electrophorus left the plate with a positive charge.
   a. How could you use the positively charged electrophorus to place a positive charge on the electroscope?

_____

_____

_____

   b. How could you use the positively charged electrophorus to place a negative charge on the electroscope?

_____

_____

_____

| **CONCEPTUAL PHYSICS** | **Demonstration** |

Charging Ahead

Purpose
To observe the effects and behavior of static electricity

Apparatus
2 new balloons
Van de Graaff generator
several pie tins (small: 5" diameter preferred)
several Styrofoam bowls
bubble-making materials (see note to right)

Going Further
wooden matches

Bubble solution and a wand are OK, but a bubble gun is easier to work with and more fun!

Discussion
Scuff your feet across a rug and reach for a doorknob and zap—electric shock! The electrical charge that makes up the spark can be several thousand volts, which is why technicians have to be so careful when working with tiny circuits such as those in computer chips!

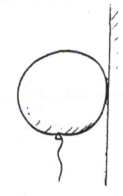

Procedure
Step 1: Blow up a balloon. After stroking it against your hair, place it near some small pieces of Styrofoam or puffed rice. Then place the balloon against the wall where it will "stick," as shown on the right. On the drawing, sketch the arrangement of some sample charges on the balloon and on the wall.

Step 2: Blow up a second balloon. Rub both balloons against your hair. Do they attract or repel each other?

_____

_____

Step 3: Stack several pie tins on the dome of the Van de Graaff generator. Turn the generator on. What happens and why?

_____

_____

_____

_____

Step 4: Turn the generator off and discharge it with the discharge ball or by touching it with your knuckle. Stack several Styrofoam bowls in the generator and turn the generator on. What happens and why?

_____

_____

_____

Step 5: With the Van de Graaff generator off and discharged, blow some bubbles toward it. Observe the behavior of the bubbles. Then turn the generator on and blow bubbles toward it again. Watch carefully. What happens and why?

_____

_____

_____

Step 6: Stand on an isolation stand (or rubber mat) next to a discharged Van de Graaff generator. Place one hand on the conducting sphere on top of the generator and have your partner switch on the generator motor. Shake your head as the generator charges up. What do you experience?

_____

_____

_____

Summing Up
Which of the demonstrations in this activity are better explained by the principle that like charges repel and opposites attract? Which are better explained in terms of the differences between conductors and insulators?

_____

_____

Going Further
Light a wooden match and move it near a charged generator. What happens and why?

_____

_____

| CONCEPTUAL PHYSICS | Tech Lab |

Electric Field Hockey

Purpose

To use a series of simulated game-like challenges to gain a sense of the vector nature of the electric field and the consequence of the inverse square law that governs electrostatic force

Apparatus

computer
PhET simulation: "Electric Field Hockey" (available at http://phet.colorado.edu)

Discussion

In the space around a charged object is an ***aura*** generally described as an electric field. The electric field is a vector quantity, so the fields surrounding multiple charges add vectorially. If they have the same direction, the fields add (Figure 1.a), and if they have opposite directions, they subtract and may even cancel each other (Figure 1.b). And if the directions are at angles to each other, their effects combine and act at a new angle (Figure 1.c).

Figure 1.a. Same direction—fields add

Figure 1.b. Opposite direction—fields cancel

Figure 1.c. Different directions—fields add to a different direction

Software

In "Electric Field Hockey," a player guides a charged puck into the goal (bracket). The player must strategically place positive and negative charges on the "rink" so that when the puck is released, it will be propelled by electric forces into the goal. Clicking the on-screen "Start" button sets the puck free. The charges placed on the rink are fixed in place and will not move.

Procedure

Step 1: Turn on the computer and allow it to complete its start-up cycle.

Step 2: Locate and start the PhET "Electric Field Hockey" simulation. Ask your instructor for assistance if you have difficulty.

Step 3: From the selections available near the bottom of the simulation window,
a.   leave "Puck is positive" selected.
b.   activate the "Trace." Leave the other checkboxes unchecked.
c.   set the mass slider control to 100. (Doing so makes it easier to follow the action on the screen.)
d.   leave the simulation in "Practice" mode for now.
e.   Complete the practice exercises that follow.

Continue until your lab time runs out or you win the prized and highly coveted Coulomb Cup (completion of Level 3).

Optional (if using the activity as a time-limited competition): When you score a prescribed goal, leave the trace on the screen and call the instructor over for a verification/stamp. ***After*** the group's papers have been stamped, click the on-screen "Clear" button and move to the next exercise.

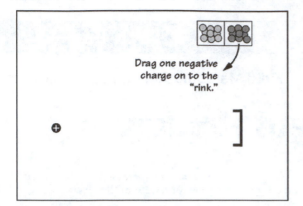

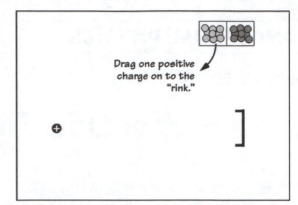

PRACTICE 1: Use one negative charge to score a goal. Sketch the location of your goal-scoring charge placement.

PRACTICE 2: Use one positive charge to score a goal. Don't forget the sketch on this and all the rest of the configurations.

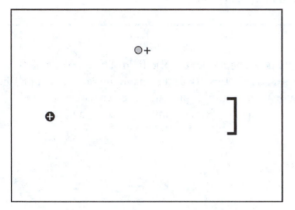

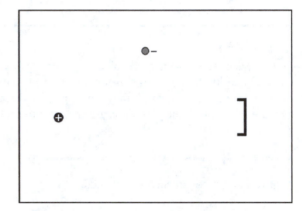

PRACTICE 3: Place one positive charge as shown. (It forms an isosceles triangle with the puck and goal.) Use one additional positive charge to score a goal.

PRACTICE 4: Place one negative charge as shown. (It forms an isosceles triangle with the puck and goal.) Use one additional positive charge to score a goal.

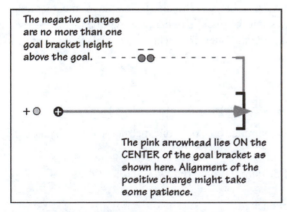

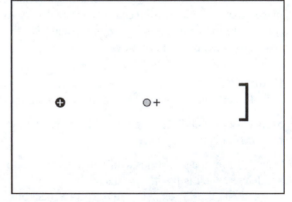

PRACTICE 5: The Double Negative: Place two negative charges and one positive charge as shown. The positive charge produces a force vector on the puck that extends <u>exactly to the center of the goal</u>. Use one additional negative charge to score.

PRACTICE 6: Positively Blocked: Place one positive charge as shown. (It's the midpoint on a straight line connecting the puck and goal.) Use two additional positive charges to score a goal. Note: this one requires patience and is sensitive to small changes!

Details and Hints

1. Do not place charges closer than 1 centimeter from one another. Placement of a charge closer than that to the puck is acceptable. Do not place any charges inside the goal bracket.

2. The puck must stay "in bounds" (on the visible portion of the rink) from beginning to end.

3. To stop the simulation and reset the puck, click the on-screen "Reset" button. To remove all placed charges on the screen, click the on-screen "Clear" button.

4. If your experimental configuration does not score a goal within 10 seconds, click the on-screen "Reset" button. A repeating, cyclical pattern can be mesmerizing to watch, but the clock is ticking.

5. If the puck wanders off the rink, click the on-screen "Reset" button.

6. Notice that small changes in the charge configuration can have large consequences for the resulting path taken by the puck.

7. After scoring a goal, activate the field display (click the on-screen "Field" checkbox) and replay the goal. You may leave the field activated or deactivate it before proceeding on to the next configuration.

8. If the program crashes, click "Ignore" in the crash alert box. Restart the simulation from the web page (as you did to start the simulation the first time).

Step 4: Once you have completed the practice exercises above, select Difficulty 1 from the choices at the bottom of the "Electric Field Hockey" window. As you proceed through the following exercises, pay particular attention to the difficulty level and stated restrictions, if any.

Difficulty 1 – Freestyle: Use as many charges as you like to score a goal. No sketch needed. _____

Difficulty 2 – Freestyle: Use as many charges as you like to score a goal. No sketch needed. _____

Difficulty 1 – Limited: Use only two positive charges to score a goal. Sketch your solution.

Difficulty 1 – Q-tip Dipole –/+: Arrange charges as shown in Figure 2. Use as many additional charges as you like. Sketch your solution.

Figure 2

Electric Field Hockey

Difficulty 2 – Exclamation Double Point Limited: Arrange the charges as shown in Figure 3. Then use no more than two positive and two negative charges to score a goal. Sketch your solution.

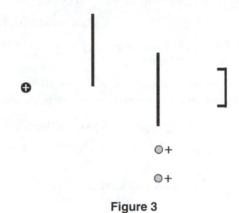

Figure 3

Going Further
Difficulty 3 – Freestyle: Use as many charges as you like to score a goal. No sketch needed. _____

Summing Up
1. If you could use only one charge to play goalie (to block goals from being made), what kind of charge would you use and where would you place it? Draw a diagram and explain your answer.

   _____

   _____

   _____

   _____

2. Show the continuation of the path that will be followed by the positive puck in each case below.

   a.

   b.

   c.

CONCEPTUAL PHYSICS	Activity

Chapter 23: Electric Current   Battery Basics and a Basic Battery Puzzle

The Lemon Electric

Purpose
To use the voltmeter function of a digital multimeter to find the relationship between the voltage of a battery and its size and to create a battery using simple materials

Apparatus
digital multimeter (DMM)   C- or D-cell alkaline battery
ignition dry cell (No. 6) battery (if available)   small alkaline battery (N- or AAAA-cell)
lemon half or wedge (or equivalent)   2 galvanized nails
plastic plate   access to water and paper towel (for clean-up)

Going Further
pennies   aluminum foil

Discussion
As more and more devices are electronic, more and more sources of energy are required to power them. The most common of these is the battery, composed of a combination of cells. In this activity, you'll consider the common battery and a bit of its history.

Procedure

PART A: SIZE AND VOLTAGE
Step 1: Arrange the multimeter to measure volts (DC). Set the dial to "V" with a range of perhaps 2 V, 20 V, or 200 V. Connect the black multimeter lead to the "common" (often black) jack. Connect the red lead to the red jack labeled with a V.

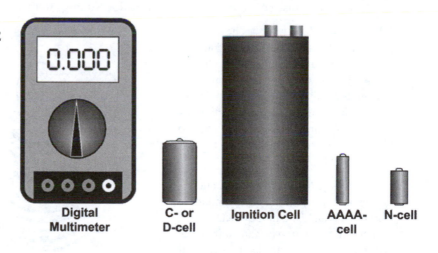

Digital Multimeter   C- or D-cell   Ignition Cell   AAAA-cell   N-cell

Step 2: a.   Measure the voltage of the C- or D-cell.

V = _____

b.   The large No. 6 ignition battery has a volume more than 20 times that of the C- or D-cell battery. Measure and record the voltage of the large battery.

V = _____

c.   An N-cell battery has about 1/8 the volume of the C- or D-cell. An AAAA-cell has about 1/20 the volume of a C- or D-cell battery. Measure and record the voltage of the small battery.

V = _____

Step 3: Consider your observations in the previous step. What relationship—if any—is apparent between the size of a battery and its voltage?

___

PART B: BACKGROUND
Read the following to learn the basic nature of the battery.

SCIENCE GETS THE BOOT FROM ITALY
When Galileo was sentenced to house arrest for his "crimes" against the Roman Catholic Church, the Scientific Revolution fled Italy for the more hospitable regions of France and England. Later, in England, Isaac Newton advanced the understanding and mathematical description of the dynamics of motion, gravity, and optics. In British America, Benjamin Franklin advanced the understanding of static electricity, and in France, Charles Coulomb applied Newtonian mathematical rigor to the description of electrostatic forces.

LUIGI AND ALESSANDRO: SCIENCE RETURNS TO ITALY
A century and a half after Galileo's death, something of scientific importance was to develop in Italy. During the 1780s, biologist Luigi Galvani performed experiments at the University of Bologna involving electrical impulses and frogs. It had been found that a charge applied to the spinal cord of a frog could generate muscular spasms throughout its body. Charges could make frog legs jump, even if the legs were no longer attached to a frog. While cutting a frog leg, Galvani's steel scalpel touched a brass hook that was holding the leg in place. The leg twitched. Further experiments confirmed this effect. Galvani was convinced that he was seeing the effects of what he called "animal electricity," the life force within the muscles of the frog.

At the University of Pavia, Galvani's colleague Alessandro Volta was able to reproduce the results, but was skeptical of Galvani's explanation. Volta, a former high school physics teacher, found that it was the presence of two dissimilar metals—not the frog leg—that was critical. In 1800, after extensive experimentation, he developed the voltaic pile. The original voltaic pile consisted of a stack of zinc and silver discs and between alternate discs, a piece of cardboard that had been soaked in saltwater. A wire connecting the bottom zinc disc to the top silver disc could produce repeated sparks. No frogs were harmed in the production of a voltaic pile.

Before the voltaic pile was developed, sparks had to be generated by friction. That involved work. When the charge was released, another spark could be generated only by more frictional work. The voltaic pile provided a continuous source of charge flow. No work had to be done! Volta's pile is widely regarded as the first battery.*

The terms *galvanic, galvanize,* and *galvanometer* (among others) honor Luigi Galvani. The unit of electric potential, the *volt,* was named in honor of Alessandro Volta. Electric potential, itself, is commonly referred to as *voltage.*

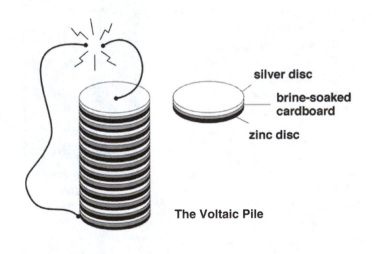

silver disc
brine-soaked cardboard
zinc disc

The Voltaic Pile

___

*An item found in an archeological excavation beneath the city of Baghdad consisted of a ceramic vessel enclosing a copper cup, which, in turn, enclosed a smaller silver cup. This reveals that a battery may have existed in ancient Persia. Unfortunately, it is not known why it was constructed or what it was used for. However, these details are known in the case of the voltaic pile.

What were the *critical ingredient(s)* of Galvani's and Volta's batteries?

_____

PART C: THE LEMON BATTERY

Step 1: Use the lemon, nails, *and your resourcefulness and creativity* to design a battery that will register 0.5 V or more. (Don't worry if the reading is negative; what's important is that the magnitude is 0.5 or greater.) When you think you have the correct design, ask the instructor to let you make a voltage measurement using the digital multimeter.

When using the multimeter, be sure to press the leads firmly against the elements of your battery. A gentle touch won't ensure the complete connection you'll need to register an accurate reading.

Always keep in mind the critical elements of the battery!

Step 2: When you achieve success,

a. record your qualifying voltage:_____.

b. describe your successful design. Where did your resourcefulness and creativity come into play?

_____

_____

Step 3: Draw two diagrams in the space below. Draw one for a configuration that doesn't work and one for a configuration that does work. Title each diagram appropriately.

Going Further

1. Given access to other materials (such as aluminum foil, pennies, and paper towel) in addition to your lemon, can you create a voltaic pile?

a. How do you properly create a voltaic pile?

People with dental fillings can make a battery in their mouth if they accidentally bite aluminum foil. It's painful, so they don't tend do it on purpose!

_____

_____

_____

b. What's the highest voltage you can attain?

_____

CONCEPTUAL PHYSICS	**Experiment**

Chapter 23: Electric Current   Connecting Meters and Determining Resistance

Ohm, Ohm on the Range

Purpose
In this experiment, you will arrange a simple circuit involving a power source and a resistor. You will attach an ammeter and a voltmeter to the circuit. You will measure corresponding values of current and voltage in the circuit. You will then interpret your observations to find the relationship between current, voltage, and resistance.

Apparatus
variable DC power supply (0–6 V)
2 power resistors with different resistances (values between 3 Ω and 10 Ω recommended)
power resistor with unknown resistance (for the "Going Further" section of experiment)
miniature lightbulb in socket (14.4-V flashlight bulb recommended)
DC ammeter (0–1 A analog recommended)
DC voltmeter (0–10 V analog recommended)
5 connecting wires
graph paper

Discussion
The current, voltage, and resistance in an electric circuit have a very specific relationship to one another. Designers of electric circuits must take this relationship into account, or their circuits will fail. This relationship is as important and fundamental in electricity as Newton's second law of motion is in mechanics. In this experiment, you will determine this relationship.

Procedure
PART A: CONNECTING THE METERS
Step 1: With the power supply turned down to zero, arrange a simple circuit using the power supply, the miniature bulb, and two connecting wires as shown in Figure 1. If your power supply has "AC" terminals, do not use them; connect only to the "DC" terminals.

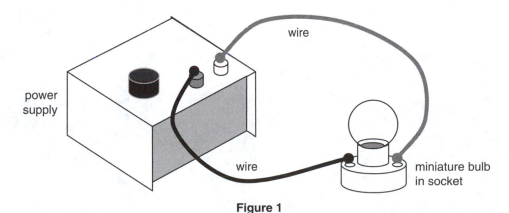

Figure 1

Step 2: Verify that the circuit is working properly by slowly turning the knob on the power supply to increase the power to the circuit. The bulb should begin to glow and increase in brightness as power is increased. If the circuit doesn't work, make adjustments so that it does. Ask your instructor for assistance if necessary.

Step 3: Turn the power down to zero.

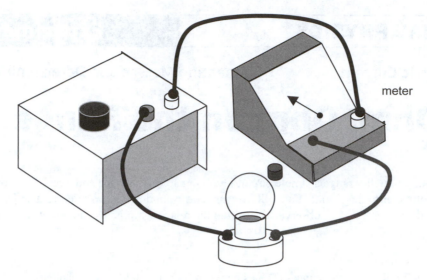

Figure 2. Meter connected in series

Step 4: Connect the ammeter in series as shown in Figure 2. If the meter has multiple positive (+) terminals, use the one labeled "1." Slowly increase the power to the circuit and observe the ammeter and the bulb. If the needle on your ammeter ever moves in the wrong direction (that is, tries to "go negative"), reverse the connections *on the meter* and try again.

When the connections are correct, what do you observe? Discuss the light—if any—from the bulb and the movement—if any—of the needle on the meter. You may get light with no meter activity, meter activity with no light, or both light and meter activity. If you get *no* light and *no* meter activity, ask your instructor for assistance.

_____

_____

_____

Step 5: Turn the power back down to zero. Disconnect the ammeter and restore the circuit to its original configuration (shown in Figure 1).

Step 6: Connect the ammeter in parallel as shown in Figure 3. Slowly increase the power to the circuit and observe the ammeter and the bulb.

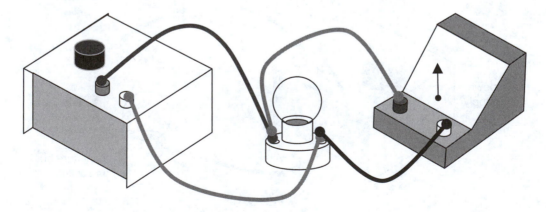

Figure 3. Meter connected in parallel

What do you observe? Discuss the light—if any—from the bulb and the movement—if any—of the needle on the meter. (Remember to switch the connections to the meter if the needle appears to move backward.)

_____

_____

_____

Step 7: Turn the power back down to zero. Disconnect the ammeter and restore the circuit to its original configuration (shown in Figure 1).

Step 8: Connect the voltmeter in series as shown in Figure 2. Slowly increase the power to the circuit and observe the voltmeter and the bulb. If the needle on your voltmeter ever moves in the wrong direction (tries to "go negative"), reverse the connections and try again.

When the connections are correct, what do you observe?

_____

_____

_____

Step 9: Turn the power back down to zero. Disconnect the voltmeter and restore the circuit to its original configuration (shown in Figure 1).

Step 10: Connect the voltmeter in parallel as shown in Figure 3. Slowly increase the power to the circuit and observe the voltmeter and the bulb.

What do you observe?

_____

_____

_____

Step 11: Synthesize your findings about connecting the meters.

a. When the ammeter is connected correctly, the circuit behaves as it did when no meters were connected (increasing the power increases the brightness of the bulb). When the power is increased, the ammeter shows increased current in the circuit. When the ammeter is connected incorrectly, the bulb remains dim or does not light at all although the ammeter shows significant current.

The correct method is to connect the ammeter to the circuit in (select one)
____series.   ____parallel.

b. When the voltmeter is connected correctly, the circuit behaves as it did when no meters were connected (increasing the power increases the brightness of the bulb). When the power is increased, the voltmeter shows increased voltage in the circuit. When the voltmeter is connected incorrectly, the bulb remains dim or does not light at all although the voltmeter shows significant voltage.

The correct method is to connect the voltmeter to the circuit in (select one)
____series.   ____parallel.

Step 12: Connect a circuit that includes the bulb, the ammeter, and the voltmeter. Use no more than five wires. Connect the ammeter and voltmeter correctly, based on your findings. Make no more than two connections to the power supply. Sketch a diagram in the space below to show how the wires connect the various circuit elements. Do not cross the lines representing wires in your sketch. Label the power supply, bulb, ammeter, and voltmeter.

Step 13: Verify that the circuit is working correctly. When the power is increased, the bulb should get brighter and *both* meter readings should increase. Show your instructor your working circuit.

PART B: COLLECTING DATA

Step 1: Turn the power down to zero. Replace the bulb with one of the known resistors. (Remove the bulb and socket from the circuit; connect the resistor in its place.)

Step 2: Record the Ω value of the known resistor in the heading of the column for the first known resistor on the data table below. The Ω value is written on the resistor, and is between 3 and 10.

Step 3: Increase the power until the current indicated on the ammeter is 0.10 A.

Step 4: Record the corresponding voltmeter reading in the appropriate space on the data table.

Step 5: Repeat Steps 3 and 4 for current values of 0.20 A through 0.60 A (or the highest value you can get on the ammeter in case you can't get all the way up to 0.60 A).

Step 6: Turn the power down to zero. Replace the first resistor with the second resistor.

Step 7: Repeat Steps 2 through 5 to complete the data table for the second resistor.

Data Table

Current I (amperes)	1st Known Resistor: _____ Ω Voltage V (volts)	2nd Known Resistor: _____ Ω Voltage V (volts)	
0	0	0	
0.10			
0.20			
0.30			
0.40			
0.50			
0.60			

Step 8: On your graph paper, plot graphs of voltage versus current for both resistors. Plot both data sets on one set of axes. Voltage will be the vertical axis; current will be the horizontal axis. Scale the graph to accommodate all your data points. Label each axis as to its quantity and units of measurement.

Step 9: Make a line of best fit for each data set plotted on the graph.

Step 10: Determine the slope of each best-fit line.

What is the slope of each best-fit line? Identify each slope by its corresponding resistor value. Don't forget to include the correct units for each slope value.

_____-Ω resistor best-fit line slope = _____

_____-Ω resistor best-fit line slope = _____

Observe the similarity between the values of the slope and the values of the corresponding resistance.

Going Further

Obtain a power resistor with an unknown resistance from your instructor. Using the techniques of this experiment, determine the resistance of the resistor. Record your data and calculations in the space below.

Once you determine the resistance of the unknown resistor, ask your instructor for the accepted value. Record that value and calculate the percent error in your value.

Ohm, Ohm on the Range

Summing Up

1. Which mathematical expression shows the correct relationship between current, voltage, and resistance?

   ___R = IV   ___R = I/V   ___R = V/I

   The correct answer is one form of the equation known as Ohm's Law.

2. Suppose voltage versus current data were taken for two devices, A and B, and the results were plotted to form the best-fit lines shown in Figure 4. Which device has the greater resistance? How do you know?

   _____

   _____

   _____

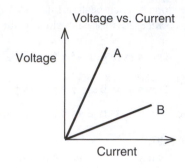

Voltage vs. Current

Fig. 4

3. Suppose the voltage and current data for electrical device C produced graph C shown in Figure 5. How does the resistance of device C change as the current increases? Explain.

   _____

   _____

   _____

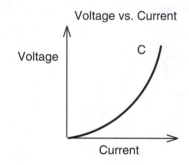

Voltage vs. Current

Fig. 5

4. Suppose the voltage and current data for electrical device D produced graph D shown in Figure 6. How does the resistance of device D change as the current increases? Explain.

   _____

   _____

   _____

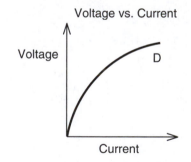

Voltage vs. Current

Fig. 6

CONCEPTUAL PHYSICS   **Tech Lab**

Ohm, Ohm on the Digital Range

Purpose
To use computer-interfaced voltage and current sensors to determine the resistance of various resistors

Apparatus
computer with data collection and analysis software
interface device
current sensor
voltage sensor
variable DC power supply (0–6 V): **_regulated, low ripple_**
2 power resistors with different resistances (values between 3 Ω and 10 Ω recommended)
power resistor with unknown resistance
miniature lightbulb in socket (14.4-V flashlight bulb recommended)
connecting wires
access to a printer

Discussion
In "Ohm, Ohm on the Range," you learned how to connect an ammeter and a voltmeter to a resistor in an electric circuit. You then collected current and voltage data. Analysis of that data allowed you to determine the resistance of the resistor. In "Ohm, Ohm on the Digital Range," you'll do much the same using a computer and sensors. The results are typically more accurate and precise!

Procedure

SETUP
Step 1: Turn on the computer and allow it to complete its start-up cycle.

Step 2: Examine your power resistors with known values.

a.   What is the resistance of your low-resistance resistor? _____ Ω.

b.   What is the resistance of your high-resistance resistor? _____ Ω.

Step 3: Examine your sensors.

a.   What is the maximum current allowed by your current sensor? _____ A.

b.   What is the maximum voltage allowed by your voltage sensor? _____ V.

Step 4: Connect a simple circuit using the power supply, low-resistance resistor, and two connecting wires.

Step 5: Replace one of the connecting wires with the two leads of the current sensor so that the current sensor is connected in **_series_** with the resistor.

Step 6: Connect the leads of the voltage sensor in parallel with the resistor.

Step 7: Connect the current and voltage sensors to the computer. The sensors will connect to the interface device, and the interface(s) will connect to the computer.

Step 8: The computer may react to the connection of the sensors with a prompt, asking you what you'd like to do. Launch your data collection software.

Step 9: In the data collection program, prepare a blank experiment document with displays for current and voltage values.

Step 10: Activate the program's data-monitoring feature. This feature allows users to see data values without collecting them for analysis.

Step 11: Turn the power supply on and slowly increase the power. Keep the current and voltage value less than the sensor's maxima (as noted in Step 3).

Both current and voltage readings should have *positive* values. If they do not, turn the power supply back down and off, then switch the sensor connections as needed.

When the connections are correct, the current and voltage readings will be positive and will increase when the power is increased.

Step 12: Deactivate the data-monitoring feature.

PART A: PREPARE A CURRENT VS. VOLTAGE GRAPH
Data collection and analysis programs vary in their functionality. The process described below is general. Your instructor will provide further guidance specific to the software you are using.

Step 1: You may remove any digital displays of current and voltage values, software permitting.

Step 2: Prepare a blank graph on which corresponding current vs. voltage values will be plotted.
a.   The graph will have current on the vertical axis and voltage on the horizontal axis.
b.   If there is an option for the software to "connect data points," turn it off; you are producing a "scatter graph."
c.   Maximize the graph window so that you will be able to see details as clearly as possible

Step 3: Before collecting data, save your experiment file to the computer's storage drive. Your instructor will provide the specifics for an appropriate location and file name.

PART B: COLLECTING THE DATA
Step 1: With the power of the power supply turned down or off, initiate data sampling. (For example, click the on-screen "Start" button.)

Step 2: *Slowly* increase the power, but keep the current and voltage below the limits of the sensors. When you reach the maximum power allowable, *slowly* decrease the power to zero again. The process should take at least 30 seconds. It should also leave a fairly even track of data points on the graph.

If you exceed the current limit of the sensor, an audible alarm may sound. Turn the power off, stop the data run, delete the data run, and return to Part B, Step 1.

Step 3: When you have turned the power all the way down, discontinue data sampling.

Step 4: Resave the experiment file with the new data set included.

Step 5: Disconnect the low-resistance resistor and replace it with the high-resistance resistor.

Step 6: Repeat Part B, Steps 1–4, with the high-resistance resistor in place.

Step 7: Disconnect the high-resistance resistor and replace it with the unknown resistor.

Step 8: Repeat Part B, Steps 1–4, with the unknown resistor in place.

PART C: LABELING THE DATA SETS
Step 1: If there is an option to "auto-scale" or "scale to fit" the plot on the graph, do so now.

Step 2: Label each data set to identify the resistor being observed. That is, if a data set represented corresponding current and voltage values for a 5-Ω resistor, label the data set as "Five Ohms." It's better to write out the label in *words* rather than numbers.

Step 3: Resave the experiment file with the data sets labeled.

PART D: DETERMINING THE SLOPE OF EACH PLOT
Step 1: Select one of your data sets and use the program's feature that allows you to impose a linear fit. Doing so should plot a line of best fit on the plot and give you access to the slope value.

Step 2: Repeat this process for the other data sets.

Step 3: Resave the experiment file with the slope information now displayed.

PART E: PRINTING THE GRAPH
Step 1: The graph may now appear cluttered because it shows data, lines of best fit, and slope information. Arrange the moveable items on the graph so that information boxes don't cover data points. (They *may* cover extended lines of best fit.)

Step 2: Secure approval from your instructor.

Step 3: Make sure you are connected (with or without wires) to a printer. Print one copy of the graph for each member of your lab group.

Summing Up

1. What is the slope given by the linear fit for the two known resistors? Don't forget to include the correct units for each slope value.

   Low: _____– Ω resistor best-fit line slope = _____

   High: _____– Ω resistor best-fit line slope = _____

2. a. What is your experimental value for the resistance of your unknown resistor?

   _____

   b. What is the accepted value for the resistance of your unknown resistor?

   _____

   c. What is the percent error in your experimental value of the unknown resistor?

   _____

Going Further

1. Using the techniques of this tech lab, complete a trial using the 14.4 V bulb as the resistor. The greatest challenge here is getting all the connections made. Don't forget to resave the file to include the new data set. Do *not* print the new graph.

2. Sketch and describe the resulting plot. Discuss similarities to and differences from the plots of the known resistors.

   _____

   _____

   _____

3. Interpret the resulting plot. What does it tell you about the resistance of the bulb? How is the bulb different from the resistors?

   _____

   _____

Science is the great antidote to the poison of superstition.

Adam Smith

CONCEPTUAL PHYSICS	Tech Lab

Resistance Is Not Futile

Purpose
To find the resistance of several different wires to determine how the composition, length, and diameter of a wire affect its electrical resistance

Apparatus
computer with data collection and analysis software
interface device
current sensor
voltage sensor
connecting wires
resistance spools
variable DC power supply (0–6 V): *regulated, low ripple*
miniature lightbulb in socket (14.4-V flashlight bulb recommended)
access to the Internet (or printed AWG information)
access to a printer

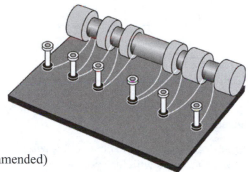

Discussion
In "Ohm, Ohm on the Digital Range," you learned how to use a computer, sensors, and data analysis software to determine the resistance of a resistor. You will use the same technique in this tech lab.

This time, the power resistors have been replaced by resistance spools. Resistance spools are made of wires with known characteristics. Since the wires are very long, they've been wrapped around cylinders to form spools. Each spool is identified by its composition, gauge (thickness), and length.

By measuring the resistance of each spool, you can find the connection between a wire's electrical resistance and its length, diameter, and composition.

After completing the tech lab, you should be able rank the following copper cylinders from lowest to highest resistance.

| Short and Narrow | Long and Narrow | Short and Wide | Long and Wide |

Procedure
SETUP
Step 1: Turn on the computer and allow it to complete its start-up cycle.

Step 2: Examine your resistance spools and record the known characteristics of each one on the data table. (The table may have more rows than you need; don't worry if you leave some blank.) Leave the diameter, area, and resistance columns blank for now.

Step 3: Examine your sensors.

a. What is the maximum current allowed by your current sensor? _____ A.

b. What is the maximum voltage allowed by your voltage sensor? _____ V.

Data Table

Spool	Composition	Length (m)	Gauge Number	Diameter (mm)	Area (mm²)	Resistance (Ω)
1						
2						
3						
4						
5						
6						

Step 4: Connect a simple circuit using the power supply, the first spool listed on the data table, and two connecting wires.

Step 5: Replace one of the connecting wires with the two leads of the current sensor so that the current sensor is connected in *series* with the resistor.

Step 6: Connect the leads of the voltage sensor in parallel with the resistor.

Step 7: Connect the current and voltage sensors to the computer. The sensors will connect to the interface device, and the interface(s) will connect to the computer.

Step 8: The computer may react to the connection of the sensors with a prompt, asking you what you'd like to do. Launch your data collection software.

Step 9: In the data collection program, prepare a blank experiment document with displays for current and voltage values.

Step 10: Activate the program's data-monitoring feature. This feature allows users to see data values without collecting them for analysis.

Step 11: Turn the power supply on and slowly increase the power. Keep the current and voltage value less than the sensor's maxima (as noted in Step 3).

Both current and voltage readings should have *positive* values. If they do not, turn the power supply back down and off, then switch the sensor connections as needed.

When the connections are correct, the current and voltage readings will be positive and will increase when the power is increased.

Step 12: Deactivate the data-monitoring feature.

PART A: PREPARE A CURRENT VS. VOLTAGE GRAPH
Data collection and analysis programs vary in their functionality. The process described below is general. Your instructor will provide further guidance specific to the software you are using.

Step 1: You may remove any digital displays of current and voltage values, software permitting.

Step 2: Prepare a blank graph on which corresponding current vs. voltage values will be plotted.
a. The graph will have current on the vertical axis and voltage on the horizontal axis.
b. If there is an option for the software to "connect data points," turn it off; you are producing a "scatter graph."
c. Maximize the graph window so that you will be able to see details as clearly as possible

Step 3: Before collecting data, save your experiment file to the computer's storage drive. Your instructor will provide the specifics for an appropriate location and file name.

PART B: COLLECTING THE DATA
Step 1: With the power of the power supply turned down or off, initiate data sampling. (For example, click the on-screen "Start" button.)

Step 2: *Slowly* increase the power, but keep the current and voltage below the limits of the sensors. When you reach the maximum power allowable, *slowly* decrease the power to zero again. The process should take at least 30 seconds. It should also leave a fairly even track of data points on the graph.

If you exceed the current limit of the sensor, an audible alarm may sound. Turn the power off, stop the data run, delete the data run, and return to Part B, Step 1.

Step 3: When you have turned the power all the way down and discontinue data sampling.

Step 4: Resave the experiment file with the new data set included.

Step 5: Disconnect the resistance spool and connect to the next spool.

Step 6: Repeat Part B, Steps 1–4 with the remaining spools.

PART C: LABELING THE DATA SETS
Step 1: If there is an option to "auto-scale" or "scale to fit" the plot on the graph, do so now.

Step 2: Label each data set to identify the spool being observed.

Step 3: Resave the experiment file with the data sets labeled.

PART D: DETERMINING THE SLOPE OF EACH PLOT
Step 1: Select one of your data sets and use the program's feature that allows you to impose a linear fit on it. Doing so should plot a line of best fit on the plot and give you access to the slope. Record the resistance of the spool on the data table.

Step 2: Repeat this process for the other data sets.

Step 3: Resave the experiment file with the slope information now displayed.

PART E: PRINTING THE GRAPH
Step 1: The graph may now be appear cluttered since it shows data, lines of best fit, and slope information. Arrange the moveable items on the graph so that information boxes don't cover data point. (They *may* cover extended lines of best fit.)

Step 2: Secure approval from your instructor.

Step 3: Make sure you are connected (with or without wires) to a printer. Print one copy of the graph for each member of your lab group.

PART F: AMERICAN WIRE GAUGE
Step 1: Research the meaning of the AWG (gauge) numbers on the wires used in this lab. Find and record the diameter of each wire. Find or calculate the cross-sectional area of each wire. Recall that the area of a circle can be calculated from the diameter using $A = \pi d^2/2$.

Resistance Is Not Futile

Summing Up

1. What relationship—if any—is there between the length of a wire and its electrical resistance? Be specific: Is there a direct proportionality, inverse proportionality, or something else? And cite the evidence for your conclusion.

_____

_____

_____

_____

2. What relationship—if any—is there between the thickness of a wire and its electrical resistance? Be specific: Is there a direct proportionality, inverse proportionality, or something else? Does the relationship connect resistance to diameter or area? Cite the evidence for your conclusion.

_____

_____

_____

_____

3. What relationship—if any—is there between the composition of a wire and its electrical resistance? Does the material the wire is composed of make any difference in its resistance? Cite the evidence for your conclusion.

_____

_____

_____

_____

4. Rank the copper cylinders from lowest (1) to highest (4) resistance based on your findings in this tech lab.

_____.   _____.   _____.   _____.
Short and   Long and Narrow: Short and   Long and Wide:
Narrow: Wide:
Length: 1   Length: 2   Length: 1   Length: 2
Diameter: 1   Diameter: 1   Diameter: 2   Diameter: 2

CONCEPTUAL PHYSICS	Activity

Batteries and Bulbs

Purpose
To explore various arrangements of batteries and bulbs, and the effects of those arrangements on bulb brightness

Apparatus
2 C- or D-cell batteries
4 connecting wires
2 miniature bulbs (screw-base flashlight bulbs, 3-V to 6-V)
2 miniature bulb sockets

Discussion
Many devices include electronic circuitry, most of which are quite complicated. Complex circuits are made, however, from simple circuits. In this activity, you will build one of the simplest yet most useful circuits ever invented—that for lighting a lightbulb!

Procedure
Step 1: Remove the bulb from the mini socket.

a. In the space below, draw a detailed diagram of the bulb, showing the following parts of the bulb's "anatomy."
    glass bulb   *filament leads (tiny wires that lead to the filament)*
    screw base   *base contact (made of lead)*   *lead separator (glass bead)*

b. There are four parts of the bulb's anatomy that you can touch (without having to break the bulb). Two of them are made of **conducting** material (metal) and two are made of **insulating** material. List them.

    Conducting parts on the outside of the bulb: _____

    Insulating parts on the outside of the bulb: _____

Step 2: Examine the two diagrams of a working electrical circuit shown below. The diagram on the left shows pictorial representations of circuit elements. The diagram on the right shows symbolic representations. Use the symbolic representations in the steps that follow.

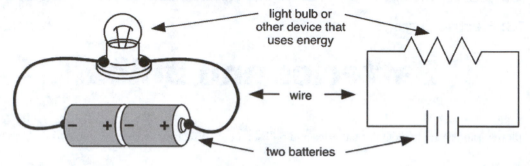

Using a bare bulb (out of its socket), one battery, and *two* wires, try lighting the bulb in as many ways as you can. On a separate sheet, sketch at least two *different* arrangements that work. Also sketch at least two arrangements that don't work. Be sure to label them as *works* or *doesn't work.*

Step 3: Using a bare bulb (out of its socket), one battery, and *one* wire, try lighting the bulb in as many ways as you can.

a.   Sketch your arrangements and note the ones that work.

b.   Is it possible to light the bulb using the battery and *no* wires? Explain.

Step 4: Connect one bulb (in its socket) to two batteries as shown in Figure 1. This arrangement is often referred to as a *simple* circuit.

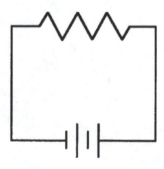

Figure 1. Simple circuit

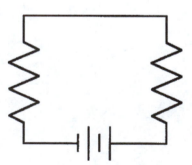

Figure 2. Series circuit

Step 5: Connect the bulbs, batteries, and wire as shown in Figure 2. When the bulbs are connected one after the other like this, the result is called a *series* circuit.

a. How does the brightness of each bulb in the series circuit compare with the brightness of the bulb in a simple circuit?

b. What happens if one of the bulbs in a series circuit is removed? (Do this by unscrewing a bulb from its socket.)

Step 6: Connect the bulbs, batteries, and wire as shown in Figure 3. When the bulbs are connected along separate paths like this, the result is called a *parallel* circuit.

a. How does the brightness of each bulb in the parallel circuit compare with the brightness of the bulb in a simple circuit?

b. What happens if one of the bulbs in a parallel circuit is removed?

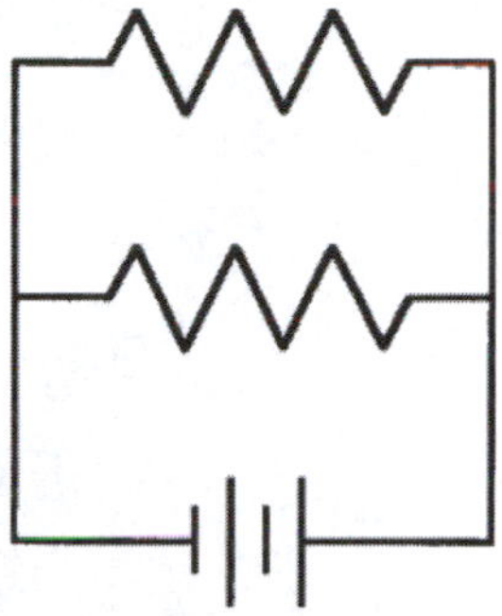

Figure 3. Parallel circuit

Summing Up

1. With what two parts of the bulb does the bulb socket make contact?

2. What do successful arrangements of batteries and bulbs have in common?

3. How do you suppose most of the circuits in your home are wired—in series or in parallel? What is your evidence?

4. How do you suppose automobile headlights are wired—in series or in parallel? What is your evidence?

Batteries and Bulbs

It is not what the scientist believes that distinguishes him, but how and why he believes it.

Bertrand Russell

| **CONCEPTUAL PHYSICS** | **Activity** |

An Open and Short Case

Purpose
To explore faulty circuits: open circuits and short circuits

Apparatus
2 D-cell batteries
5 connecting wires
2 miniature bulbs (1.5-V or 2.5-V flashlight bulbs)
2 miniature bulb sockets
DC ammeter (0–5 A)

Discussion
Electrical circuits are all around us. (We often appreciate them most when the power goes out.) Most circuits work perfectly well (when power is available). But electrical circuits *can* fail. Two common modes of circuit failure are *open circuits* and *short circuits.* In this activity, we will learn how these circuit failures are similar and how they are different.

We will be using an *ammeter* to help us with this investigation. An ammeter is a simple device used to measure electric current—the rate at which charge flows through a circuit. Current is measured in amperes ("amps").

Procedure

PART A: OPEN AND SHORT CIRCUITS
Step 1: Arrange a simple circuit using two batteries, a bulb, an ammeter, and three connecting wires as shown in Figure 1. If the circuit is working, the bulb will light and some amount of current will register on the ammeter. It should be less than 1 amp.

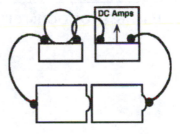

Figure 1. Simple circuit

Step 2: Predict what would happen to the simple circuit if one of the wires were disconnected at one point in the circuit. (Don't touch the circuit yet—predict first!)

a.   What will happen to the bulb and what will happen to the reading on the ammeter (compared to what happened in Step 1)?

_____

_____

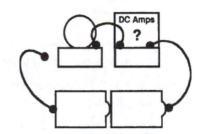

Figure 2. Open circuit

Once you've made your prediction and discussed it with your partner(s), disconnect a wire as shown in Figure 2. This is an *open circuit.*

b.   Record your observations.

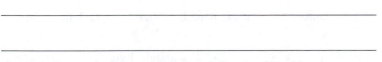

_____

_____

Step 3: Predict what would happen to the simple circuit if an additional wire were added to the circuit so as to connect the terminals of the bulb to each other as shown in Figure 3. (Don't touch the circuit yet—predict first!)

a. What will happen to the bulb and what will happen to the reading on the ammeter (compared with what happened in Step 1)?

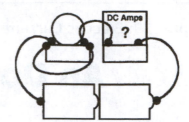

_____

_____

Once you've made your prediction and discussed it with your partner(s), add a wire as shown in Figure 3. This is a **short circuit.**

Figure 3. Short circuit

b. Record your observations.

_____

_____

Step 4: One of these circuit failures is said to have almost *no electrical resistance* and one is said to have *infinite electrical resistance.* Electrical resistance is inversely proportional to electrical current in a simple circuit.

a. Which circuit failure has no current and, therefore, infinite electrical resistance?

_____

b. Which circuit failure has a large amount of current and therefore almost no electrical resistance?

_____

PART B: SHORT-CIRCUITED SERIES CIRCUIT

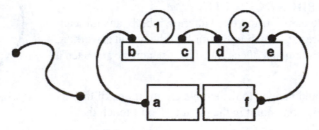

Figure 4. Series circuit and additional wire

Step 1: Arrange the series circuit shown in Figure 4. Notice that bulbs 1 and 2 light up. Notice there is an additional wire not yet in the circuit.

Step 2: Add the additional wire to the circuit, connecting point a to point b. Notice that both bulb 1 and bulb 2 remain fully lit.

Step 3: Now use the additional wire to connect point b to point c. Notice that bulb 1 goes out (or becomes much dimmer), while bulb 2 remains fully lit (or becomes brighter).

Step 4: Predict what will happen if the additional wire is used to connect other points on the circuit. Make your prediction in terms of what will happen to each of the bulbs. Will bulb 1 remain lit or go out? Will bulb 2 remain lit or go out?

Important Note: For purposes of this activity, "remaining lit" includes increased brightness, and "going out" includes significant dimming.

Remember to make predictions before making observations!

a. If point c is connected to point d, bulb 1 will __remain lit __go out and bulb 2 will __remain lit __go out.

b. If point d is connected to point e, _____

c. If point e is connected to point f, _____

d. If point f is connected to point a, _____

e. If point a is connected to point c, _____

f. If point a is connected to point d, _____

g. If point a is connected to point e, _____

h. If point b is connected to point d, _____

i. If point b is connected to point e, _____

Step 5: Observe what happens if the additional wire is used to make each of the connections.

a. When point c is connected to point d, bulb 1 __remains lit __goes out and bulb 2 __remains lit __goes out.

b. When point d is connected to point e, _____

c. When point e is connected to point f, _____

d. When point f is connected to point a, _____

e. When point a is connected to point c, _____

f. When point a is connected to point d, _____

g. When point a is connected to point e, _____

h. When point b is connected to point d, _____

i. When point b is connected to point e, _____

An Open and Short Case

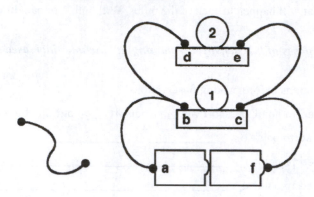

Figure 5. Parallel circuit and additional wire

Step 1: Arrange the parallel circuit shown in Figure 5. Notice that bulbs 1 and 2 light up. Notice there is an additional wire not yet in the circuit.

Step 2: Add the additional wire to the circuit, connecting point a to point b. Notice that both bulb 1 and bulb 2 remain fully lit.

Step 3: Now use the additional wire to connect point b to point c. Notice that bulb 1 goes out (or becomes much dimmer), while bulb 2 remains fully lit (or becomes brighter).

Step 4: Predict what will happen if the additional wire is used to connect other points on the circuit. Make your prediction in terms of what will happen to each of the bulbs. Will bulb 1 remain lit or go out? Will bulb 2 remain lit or go out? *For purposes of this activity, "remaining lit" includes increased brightness, and "going out" includes significant dimming.* Remember to make predictions before making observations!

a. If point c is connected to point d, bulb 1 will __remain lit __go out and bulb 2 will __remain lit __go out.

b. If point d is connected to point e, _____

c. If point e is connected to point f, _____

d. If point f is connected to point a, _____

e. If point a is connected to point c, _____

f. If point a is connected to point d, _____

g. If point a is connected to point e, _____

h. If point b is connected to point d, _____

i. If point b is connected to point e, _____

Step 5: Observe what happens if the additional wire is used to make each of the connections.

a. When point c is connected to point d, bulb 1 __remains lit __goes out and bulb 2 __remains lit __goes out.

b. When point d is connected to point e, _____

c. When point e is connected to point f, _____

d. When point f is connected to point a, _____

e. When point a is connected to point c, _____

f. When point a is connected to point d, _____

g. When point a is connected to point e, _____

h. When point b is connected to point d, _____

i. When point b is connected to point e, _____

Summing Up

1. What do open circuits and short circuits have in common?

_____

_____

2. How are open circuits and short circuits different?

_____

_____

3. Examine the cases in Parts B and C when both bulbs went out. Is there anything that *all* of those cases have in common?

_____

_____

An Open and Short Case

CONCEPTUAL PHYSICS	Activity

Be the Battery

Purpose
To provide energy to an electric circuit using your own muscle power

Apparatus
handheld generator (Genecon or equivalent)
3 miniature bulbs (6-V flashlight bulbs)
3 miniature bulb sockets
6 connecting wires

Discussion
Batteries last longer in some circuits than they do in others. They last longer when they don't have to "work" so hard. In this activity, **you** will do the work of the battery. That is, you will power a circuit using the handheld generator. You will learn which circuits are easier to power and which circuits are harder to power. And you'll gain a better appreciation for what batteries and the local power utility do for you all the time!

Procedure

Step 1: Arrange a simple circuit using the generator and a bulb in its socket as shown in Figure 1. Gently crank the handle to make the bulb light up. Take care not to crank the generator too quickly, and don't give it any sudden jerks or bursts of motion.

Figure 1. The hand-powered circuit

When the bulb is lit, how can you make it brighter? Does this require more effort on your part?

Step 2: While cranking the generator and lighting the bulb, have a partner unscrew the bulb from the socket as shown in Figure 2.

a. What happens to the cranking effort when the bulb is unscrewed from its socket?

_____

b. When the bulb is removed from the socket, is the resulting circuit an open circuit or a short circuit?

Figure 2

c. Is the electrical resistance in this kind of circuit very high or very low?

_____

Step 3: Remove the generator leads from the bulb terminals and connect them to each other as shown in Figure 3.

a. What happens to the cranking effort when the generator leads are connected to each other?

b. When the generator leads are connected directly to each other, is the resulting circuit an open circuit or a short circuit?

Figure 3

c. Is the electrical resistance in this kind of circuit very high or very low?

Step 4: Connect three bulbs in series as shown in Figure 4. Gently crank the handle to make the bulbs light up. Get a sense of how much effort is needed to power the circuit.

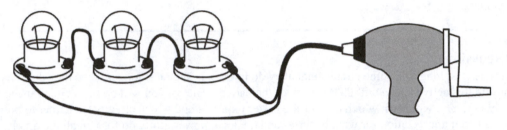

Figure 4. A hand-powered series circuit

Step 5: Connect three bulbs in parallel as shown in Figure 5. Gently crank the handle to make the bulbs light up. Get a sense of how much effort is needed to power the circuit.

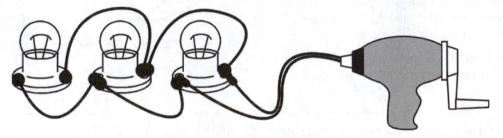

Figure 5. A hand-powered parallel circuit

Which circuit is harder to power, the series circuit or the parallel circuit?

Summing Up

1. Which types of circuits are harder to power, those having low electrical resistance or those having high electrical resistance?

2. Which arrangement of three bulbs has more electrical resistance, series or parallel? Justify your answer using observations from the activity.

3. Under which conditions will a battery run down faster, when connected to a high-resistance circuit or when connected to a low-resistance circuit?

4. Which battery would last longer, one connected to a three-bulb series circuit or one connected to a three-bulb parallel circuit? (Assume the batteries are identical and the bulbs are identical.)

Be the Battery

There is no harm in doubt and skepticism, for it is through these that new discoveries are made.

Richard Feynman

Name _____ Section _____ Date _____

| CONCEPTUAL PHYSICS | Activity |

Seeing Magnetic Fields

Purpose
To explore the patterns of magnetic fields around bar magnets in various configurations

Apparatus
3 bar magnets
iron filings and paper or a magnetic field projectual (iron filings suspended in oil encased in an acrylic envelope)

Discussion
An electric field, as we have learned, surrounds electric charges. In a similar way, a magnetic field surrounds magnets. (Magnetic fields surround other things, too, as we'll learn later.) In this activity, we'll examine the magnetic fields surrounding bar magnets. Although the fields can't be seen directly, their overall shape can be seen by their effect on iron filings.

Procedure
Step 1: Place a bar magnet on a horizontal surface such as your tabletop. Use the iron filings to see the pattern of the magnetic field.

PART A: Iron Filings and Paper Method
Cover the magnet or magnets with a sheet of paper. Then sprinkle iron filings on top of the paper. Jiggle the paper a little bit to help the iron filings find their way into the magnetic field pattern.

PART B: Projectual Method
Step 1: Mix the iron filings by rotating the projectual. Use the glass rod inside the projectual to help stir the iron filings into a fairly even distribution. Hold the projectual upside down for several seconds before placing it on the magnet or magnets. Take care not to scratch the surface of the projectual by moving it across the magnets once it is in place.

Sketch the field for a single bar magnet in Figure 1.

Step 2: Arrange two bar magnets in a line with opposite poles facing each other. Leave about 1 inch between the poles. Use the iron filings to see the pattern of the magnetic field. Sketch the field for opposite poles in Figure 2.

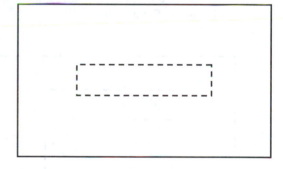

Figure 1

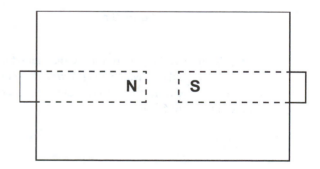

Figure 2

Step 3: Arrange two bar magnets in a line with north poles facing each other. Leave about 1 inch between the poles. Use the iron filings to reveal the magnetic field. Sketch the field for north poles in Figure 3.

Step 4: Predict the pattern for the magnetic field of two south poles facing each other as shown in Figures 4.a and 4.b. Make a *predictive* sketch in Figure 4.a. Then arrange two bar magnets in a line with south poles facing each other. Leave about 1 inch between the poles. Use the iron filings to reveal the magnetic field. Sketch the *observed* field for south poles in Figure 4.b.

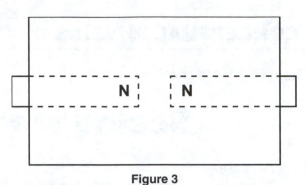

Figure 3

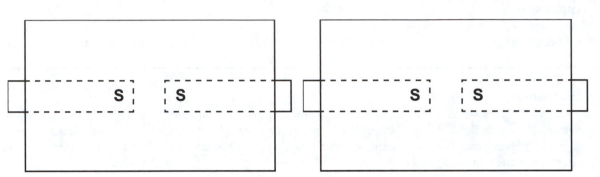

Figure 4.a   **Figure 4.b**

Step 5: Predict the pattern for the magnetic field of two bar magnets parallel to each other as shown in Figures 5.a and 5.b. Make a *predictive* sketch in Figure 5.a. Then arrange two bar magnets parallel to each other. Use the iron filings to reveal the magnetic field. Sketch the *observed* field for two magnets parallel to each other in Figure 5.b.

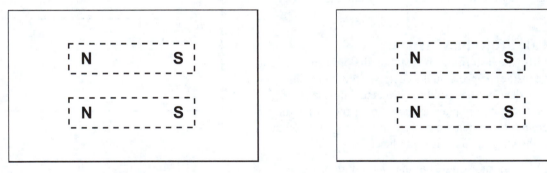

Figure 5.a   **Figure 5.b**

Step 6: Predict the pattern for the magnetic field of two bar magnets anti-parallel to each other as shown in Figures 6a. and 6.b. Make a *predictive* sketch in Figure 6.a. Then arrange two bar magnets anti-parallel to each other. Use the iron filings to reveal the magnetic field. Sketch the *observed* field for two magnets parallel to each other in Figure 6.b.

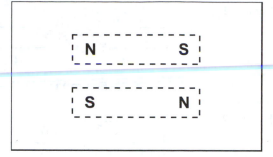

Figure 6.a

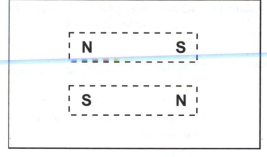

Figure 6.b

Step 7: Can you arrange three bar magnets to create the magnetic field shown in Figure 7? If so, how? Is there more than one way to do it?

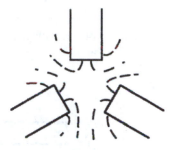

Figure 7

_____

_____

_____

a.   Does this pattern show attraction or repulsion?

Step 8: Can you arrange three bar magnets to create the magnetic field shown in Figure 8? If so, how? Is there more than one way to do it?

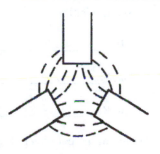

_____

_____

_____

a.   Does this pattern show attraction or repulsion?

Figure 8

Summing Up

1.   Suppose you see a magnetic field pattern as shown in Figure 9. Can you say for sure which pole is north and which pole is south?

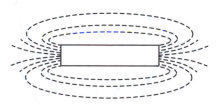

Figure 9

_____

_____

2.   Suppose you see a magnetic field pattern as shown in Figure 10. Can you say for sure which pole is north and which pole is south?

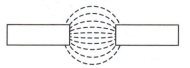

Figure 10

_____

Seeing Magnetic Fields

3. Suppose you see a magnetic field pattern as shown in Figure 11. If pole A is a north pole, what is pole B?

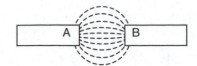

Figure 11

_____

_____

4. Suppose you see a magnetic field pattern as shown in Figure 12. If pole C is a north pole, what is pole D?

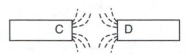

Figure 12

_____

_____

5. Suppose you see a magnetic field pattern as shown in Figure 13. If pole E is a north pole, what are poles F, G, H, I, J, K, and L?

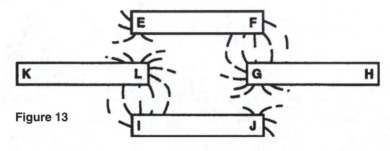

Figure 13

Pole E: North   Pole F:_____

Pole G:_____   Pole H:_____

Pole I:_____   Pole J:_____

Pole K:_____   Pole L:_____

6. Which of the patterns in Figure 14—if either—is/are possible using three bar magnets?

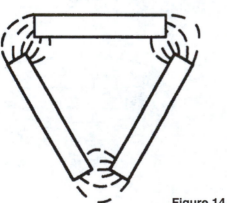

Figure 14.a

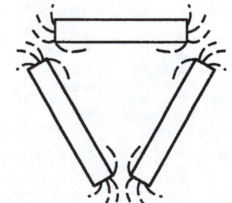

Figure 14.b

_____

CONCEPTUAL PHYSICS	**Activity**

Electric Magnetism

Purpose
To investigate the electric origin of magnetic fields

Apparatus
1 large battery (6-V lantern battery or 1.5-V ignitor battery)
4 small compasses
small platform (a discarded compact disc or equivalent flat object with a hole in the center)
support rod with base
ring clamp
connecting wires

Discussion
In "Seeing Magnetic Fields," you investigated the magnetic fields around various configurations of bar magnets. But where does the magnetic field come from? What's going on inside a bar magnet to make it magnetic? In this activity, you will discover the origin of all magnetic fields.

Procedure
PART A: CURRENT ACROSS A COMPASS
Step 1: Set a compass on your desktop and allow the needle to settle into its north–south alignment as shown in Figure 1.

Step 2: Stretch a connecting wire across the top of the compass as shown in Figure 2. Rotate the wire clockwise and counterclockwise so that you can see that the wire itself has no affect on the compass needle.

Step 3: Connect the stretched wire to the battery to form a short-circuit and again rotate the wire back and forth as shown in Figure 3. Keep the short-circuit connected for no more than 10 seconds.

What effect does the current-carrying wire have on the compass?

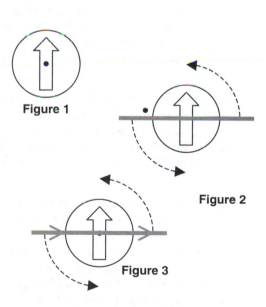

Figure 1

Figure 2

Figure 3

Step 4: Determine which is more effective in deflecting the compass needle: north–south current or east–west current. Keep in mind that short circuits must not be allowed to run more than 10 seconds and compasses must be level to work properly.

Current has the greatest effect on the compass needle when it runs (select one)

____ north–south.   ____ east–west.

Step 5: Try placing the wire *below* the compass and then running current through it.

Does the current affect the needle when the wire passes below the compass?

Step 6. Try reversing the direction of the current by reversing the connections to the battery.

What difference does reversing the direction of current have on the deflection of the needle?

PART B: CURRENT THROUGH A PLATFORM OF COMPASSES

Step 1: Arrange the apparatus as shown in Figure 4. Connecting wire passes through the center of the platform. The platform is supported by the ring clamp. The compasses are placed on the platform. Devise a method to have the connecting wire as vertical as possible as it passes through the compass platform.

Step 2: Before running any current through the wire, examine the compass needles by looking down from above the platform. Notice they all point north as indicated in Figure 5.

Step 3: Arrange to have current passing upward through the platform as shown in Figure 6. Connect the wire to the battery and tap the platform a few times. Record the new orientations of the compass needles in Figure 6.

Step 4: Reverse the direction of the current so the current passes downward through the platform. Connect the wire to the battery and tap the platform a few times. Record the new orientations of the compass needles in Figure 7.

The ability of an electric current to affect a compass needle was discovered by Hans Christian Ørsted, a Dutch high school teacher, in 1820. The observation established the connection between electricity and magnetism. We now know that *all* magnetic fields are the result of moving electric charge (even the magnetic fields of bar magnets). Ørsted's discovery stands as one of the most significant discoveries in the history of physics.

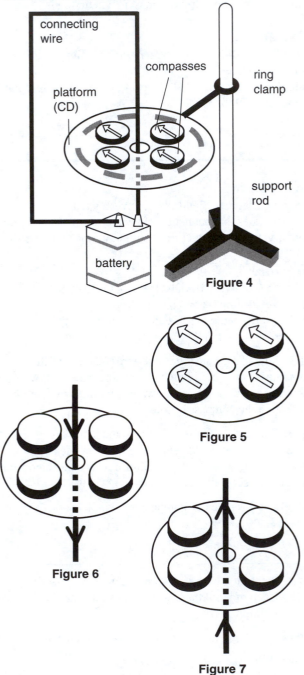

Figure 4

Figure 5

Figure 6

Figure 7

Summing Up

1. Current is passing through the center of a platform that supports four compasses. You are looking straight down at the platform. What is the direction of the current in each configuration shown below: coming toward you or going away from you?

a.

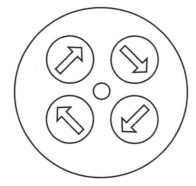

b.

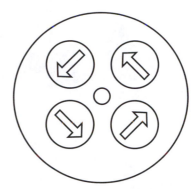

2. The direction of the magnetic field around a wire can be related to the direction of the current in the wire. If you imagine grabbing the wire with your thumb pointing in the direction of current, your fingers wrap around the wire in the direction of the magnetic field. Which hand must you use for this exercise to give the correct relationship between the direction of the current and the direction of the magnetic field: your left or right?

3. What is the source of **all** magnetic fields?

Electric Magnetism

No one really becomes a fool until they stop asking questions.

Charles Steinmetz

CONCEPTUAL PHYSICS	Activity

Motor Madness

Purpose
To investigate the principles that make electric motors possible

Apparatus
handheld generator (Genecon or equivalent)
connecting wires
about 50 cm of lead-free solder
2 collar hooks (or 2 10-cm lengths of lead-free solder)
about 30 cm of 1/4" diameter wood dowel
support rod with base and rod clamp
2 bar magnets (strong alnico magnets are recommended)

small block of wood (about 2" × 2" × 1")

2 rubber bands
2 D-cell batteries

Going Further
St. Louis Motor
6-V lantern battery

Discussion
Perhaps the most important invention of the 19th century was the electric motor. You use a motor whenever you use electric power to make something move. A motor is used to start the engine of a car. Motors are used to spin compact discs. Motors are used to move elevators up and down. A list of motor applications would go on and on. But how do motors turn electric energy into mechanical energy? Let's find out!

An electric current can exert a force on a compass needle (which is simply a small magnet). But Newton's third law of motion suggests that something else must be going on here. What is it?
Finish the statement:

If an electric current can exert a force on a magnet, then a magnet

Procedure
PART A: THE MAGNETIC SWING
Step 1: Arrange a solder "swing" by following the instructions below.

a. Make a "sandwich" with the two bar magnets and the wood block as shown in Figure 1. The magnets must have opposite poles facing each other. Secure the sandwich with the rubber band. See Figure 1.

b. Attach the support rod to the table clamp or ring stand base.

c. Attach the wood dowel to the support rod.

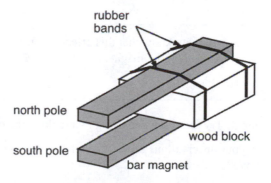

MAGNET: Sandwich a small wood block between two bar magnets. Secure the arrangement with rubber bands as shown. Note that the magnets are antiparallel: opposite poles face each other.

Figure 1

d. Place the collar hooks on the wood dowel about 10 cm apart. (You can use the short lengths of solder to make two hooks if collar hooks are not available.)

e. Bend the long length of solder into a square U-shaped "swing." Make wide hooks at the ends and hang the swing from the solder hooks on the wood dowel. *The swing must sway freely on its hooks.* See Figures 2 and 3.

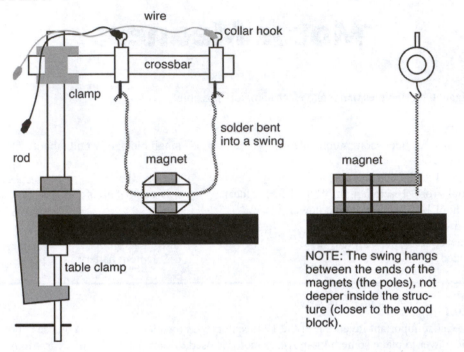

Figure 2. Front view of arrangement.   **Figure 3.** Side view of arrangement.

f. Arrange the height of the crossbar so that the bottom of the solder swing hangs between the magnets.

g. Attach the generator leads to the top leads of the hooks.

Step 2: Crank the generator one way. This will send electric current one way through the swing. Then crank the generator the other way.

a. What effect—if any—does the magnetic field of the bar magnets have on the current in the swing?

_____

b. How do you know that this effect is caused by the current's interaction with the bar magnets? (Would the swing sway if the magnets weren't there?)

_____

Step 3: Try *pumping* the swing by cranking the generator back and forth.

Do your observations of the magnetic swing confirm or contradict the prediction stated in the Discussion section above?

_____

PART B: THE SIMPLE MOTOR

Once it was found that a magnetic field could exert a force on an electric current, clever engineers designed practical ways to harness this force. They started to build electric motors. A motor transforms electric energy into mechanical energy. Some simple motors are made of coils of wire and magnets arranged so that when electric current flows through the wires, some part of the motor rotates.

The hand generator you've been using in this and other labs is, in fact, *a motor!*

Step 1: Hold the grip of the generator, but not the crank handle. Touch the two leads of the generator to opposite terminals of a single D-cell battery. What happens?

Step 2: Put two batteries together in series (in a line end-to-end) and touch the leads of the generator to opposite terminals of the arrangement. How is the result different from what happened in Step 1?

Going Further
THE ST. LOUIS MOTOR

Step 1: Obtain a St. Louis Motor and a 6-V lantern battery (or equivalent).

Step 2: Label the following parts of the St. Louis Motor in Figure 4. (Write the names of the parts, not the single-letter designations from the list!)

a. **Armature:** coiled copper wire that will carry current when the battery is connected. Current in the armature produces a magnetic field that interacts with the field magnets.
b. **Field Magnets:** permanent magnets whose fields will interact with the magnetic field in the armature.
c. **Split-Ring Commutator:** metal cylinder on the spinning axle of the motor. Notice the gaps in the metal (the "split" in the ring). The split-ring commutator directs the current through the armature so that the armature continues to spin rather than getting stuck in one position.
d. **Commutator Bridge:** the yolk that connects the brushes. The angle of the commutator bridge is adjustable.
e. **Brushes:** metal structures that touch the split-ring commutator. They are connected to the commutator bridge.
f. **Terminals:** electrical connection posts on the commutator bridge.

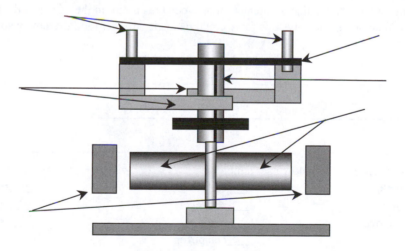

Figure 4. St. Louis Motor

SAFETY CAUTION: When the St. Louis Motor is running, the armature can spin at very high speed. Handle the motor with caution and keep it away from loose clothing, dangling jewelry, long hair, etc. Do not attempt to stop the spinning armature with fingers, pencils, etc., as injury will result.

Step 3: Connect the 6-V lantern battery to the St. Louis Motor. If the motor does not immediately spin upon connection, troubleshoot as follows:

First: Jiggle or tap the St. Louis Motor housing or base.

Second: Give the armature a spin in one direction. If that doesn't work, try spinning the armature the other way.

Third: Disconnect the battery and adjust the brushes so that they are in good contact with the split-ring commutator. Reconnect the battery and try again.

Fourth: Ask your instructor for assistance.

Step 4: Adjust the angle of the commutator bridge and record your findings.

_____

_____

Step 5: Find the optimal angle for the commutator bridge. Adjust the position of the field magnets and record your findings. If possible, remove the field magnets completely. Then try reversing the polarity of one so that north faces north or south faces south.

_____

_____

Step 6: Your battery provides a relatively constant amount of electrical power. What do you think would happen if you could increase the electrical power supplied to the motor?

_____

Summing Up

1. The magnetic swing swayed due to interaction between the current in the wire and the magnetic field of the bar magnets. What are some ways this force could be made stronger (and thereby push the swing further in or further out)?

_____

_____

_____

2. Which of the devices listed below uses a motor?

____alarm clock	____toilet	____shower
____blow dryer	____shaver	____cassette player
____CD/DVD player	____radio	____vending machine
____lightbulb	____computer	____TV
____VCR	____washing machine	____car

3. In addition to the devices identified in question 2, list two more devices that use motors.

_____

_____

| **CONCEPTUAL PHYSICS** | **Demonstration** |

Bobbing for Magnets

Purpose
In this demonstration, you will see the interaction of electricity and magnetism. Specifically, you will use a magnet to generate an electric current. And you will see how the induced current produces a magnetic field. Finally, you will see how the resulting magnetic field interacts with the original magnet that induced it.

Apparatus
demonstration galvanometer (–500 μA – 0 – +500 μA preferred)
2 bar magnets (strong alnico magnets are preferred)
2 weak springs (low force constant, approximately 2"–3" long)
2 support bases, right angle clamps, and support rods

2 crossbars	2 collar hooks
2 air core solenoids	6-V lantern battery
2 connecting wires	hand-crank generator (Genecon or equivalent)

Going Further
copper or aluminum tube
neodymium magnet (supermagnet)

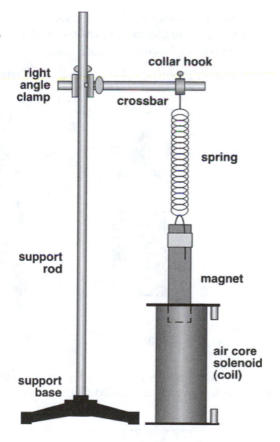

Figure 1

Discussion
Electromagnetic induction can be tricky to demonstrate and understand. This demonstration shows the effects of electromagnetic induction in a series of steps. A moving bar magnet induces a current in a conducting wire. But current in a wire induces a magnetic field. And bar magnets are affected by magnetic fields.

Procedure

SETUP
Step 1: Attach a support rod to a support stand. Use a right angle clamp to attach a crossbar to the support rod. Attach a collar hook to the crossbar.

Step 2: Attach a spring to the south pole of a bar magnet and suspend the arrangement from the collar hook.

Step 3: Feed the north pole of the suspended magnet into an upright air core solenoid (coil). Adjust the height of the crossbar so that the north pole of the magnet sits about 2 centimeters deep into the coil when the spring is at equilibrium.

PART A: ELECTROMAGNET

Step 1: Use the connecting wires to briefly connect the lantern battery to the coil. When current passes through the coil, the coil is an *electromagnet*.

a. What happens to the magnet when the connection is made?

_____

_____

b. The reaction of the bar magnet is what one might expect if the top of the solenoid became a
    __north pole   __south pole

Step 2: *Reverse the connecting wires* and briefly connect the lantern battery to the solenoid.

a. What happens to the magnet when the connection is made with reversed polarity?

_____

_____

b. The reaction of the bar magnet is what one might expect if the top of the solenoid became a
    __north pole   __south pole

Step 3: Figure 2.a shows a bar magnet dangling above a coil before the battery is connected.

a. Figure 2.b shows a bar magnet suspended above a coil while current runs through it. The current-carrying coil is an electromagnet and repels the north pole of the magnet. In the empty circles in the diagram, identify which end of the coil acts as a north pole (N) and which acts as a south pole (S).

b. Figure 2.c shows a bar magnet suspended above a coil while current runs through it. The current-carrying coil is an electromagnet and attracts the north pole of the magnet. In the empty circles in the diagram, identify which end of the coil acts as a north pole (N) and which acts as a south pole (S).

Notice that the magnetic field surrounding the current-carrying coils is shown in Figures 2.b and 2.c.

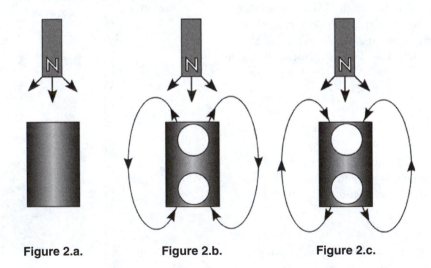

Figure 2.a.   **Figure 2.b.**   **Figure 2.c.**

Step 4: Disconnect the wires from the battery to the coil. Connect the hand-crank generator to the coil. Use the generator to move the suspended magnet and describe your observations. (Can you establish *resonance*?)

_____

_____

Bobbing for Magnets   222

Step 5: These demonstrations show that an electric current can exert a force on a permanent magnet. By Newton's third law, what must also be true?

_____

_____

PART B: INDUCTION
Step 1: Disconnect the hand-crank generator and connect the coil to the galvanometer. Move the suspended magnet in and out of the coil. Observe the corresponding motion of the galvanometer needle. How does this demonstration relate to your finding in Part A, Step 5?

_____

_____

PART C: ELECTROMAGNETIC INTERACTION
The moving magnet induces current in the coil. But current in the coil produces a magnetic field. And a magnetic field can exert a force on a magnet.

Step 1: Disconnect the galvanometer from the coil. Allow the suspended magnet to oscillate _(bob)_ in and out of the coil. With the coil disconnected, it is an open circuit and no current passes through it. With no current in the coil, there is no magnetic field around the coil.

Step 2: Connect the coil to the galvanometer and set up the oscillation again. Current now runs through the coil, and a magnetic field is produced. What effect does the magnetic field have on the motion of the bobbing magnet?
____the motion of the magnet is _increased_ by the magnetic field produced by the coil
____the motion of the magnet is _decreased_ by the magnetic field produced by the coil

PART D: DOUBLE THE FUN
Step 1: Arrange a second suspended magnet with its north pole down. Suspend it in a second coil as shown in Figure 3. Connect the two coils as shown. (If the connection posts are color-coded and on the same end of the coil, connect red to black and black to red.) If current spirals downward through the first coil, it will also spiral downward through the second coil when the coils are connected this way.

Step 2: Set one magnet into oscillation. What happens to the other magnet? Be specific in your description.

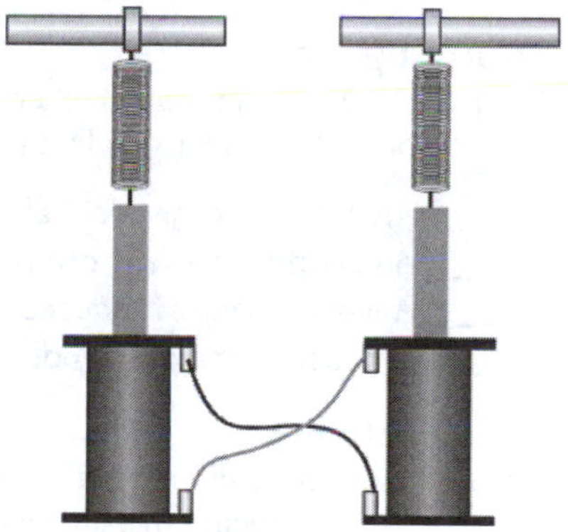

_____

_____

Figure 3

Step 3: Work out the process by correctly completing the statements that follow.

a. The bobbing magnet induces a _____ in its coil.

b. The _____ in that coil also passes through the other coil since the coils are connected.

Bobbing for Magnets

c. The _____ in the second coil produces a

_____ _____ around the

second coil.

d. The _____ _____ around the

second coil exerts a force on the bar magnet suspended into it.

e. When the first magnet is moving down, the force experienced by the second magnet is directed

_____.

f. This finding should be consistent with **Lenz's law**. One way of stating Lenz's law is that the current produced by a moving magnet creates a magnetic field that opposes the original motion of the magnet. That is to say, a downward-moving magnet produces a current that creates a magnetic field that pushes the magnet upward.

Step 4: Reverse the connection on the second coil as shown in Figure 4. What effect does this have on the oscillations and why?

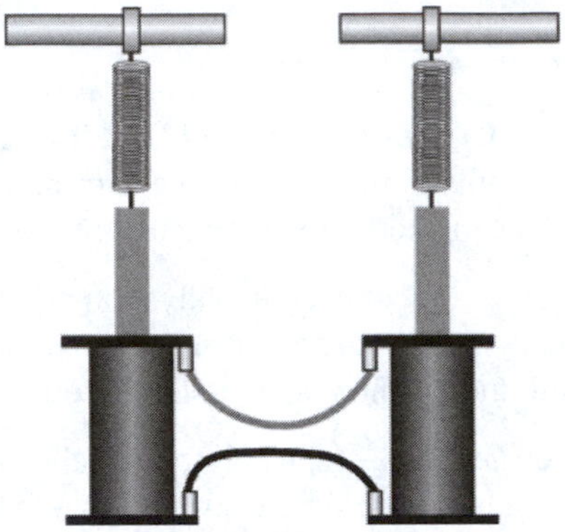

_____

_____

_____

_____

Figure 4

Summing Up

1. Label the steps in the sequence below in the order in which they occurred at the beginning of the demonstration in Part D: Double the Fun.

___An induced magnetic field exerts force on a magnet.
___An electric current creates a magnetic field.
___A moving magnet induces an electric current.
___Current in one coil produces current in the other coil.

Going Further

Drop a small neodymium magnet (**supermagnet**) into a vertical copper or aluminum pipe.
1. Observe and record the surprising result.

_____

_____

2. Use the principles discussed in "Bobbing for Magnets" to explain the surprising result.

_____

_____

_____

_____

CONCEPTUAL PHYSICS | Activity

Generator Activator

Purpose
To investigate electromagnetic induction, the principle behind electric generators

Apparatus
bar magnet (strong alnico magnet is recommended)
air core solenoid or a long wire coiled into many loops
connecting wires
galvanometer (–500 µA – 0 – +500 µA recommended)
handheld generator (Genecon or equivalent)
access to a second handheld generator

Discussion
In 1820, Hans Christian Ørsted found that electricity could create magnetism. Scientists were convinced that if electricity could create magnetism, magnetism could create electricity. Still, 11 years would pass before the induction of electricity from magnetism would be discovered and understood. The best minds of the day set out to make this widely anticipated discovery, but it was Michael Faraday who put all the pieces together. In this activity, you will use a magnet to create an electric current and see how this effect is applied in electric generators.

Procedure
PART A: ELECTROMAGNETIC INDUCTION
Step 1: Connect the galvanometer to the coil (air core solenoid or looped wire).

Step 2: Determine a method for producing current in the coil using the bar magnet.

a. Describe your findings.

_____

_____

b. Can current be produced if
   i. the coil is at rest? If so, how?

_____

   ii. the magnet is at rest? If so, how?

_____

   iii. both the coil and the magnet are at rest? If so, how?

_____

PART B: THE GENERATOR

Once it was found that a changing magnetic field could induce an electric current, clever engineers figured out practical ways to harness induction. They started to build electric generators. A generator transforms mechanical energy into electrical energy. Some simple generators are made of coils of wire and magnets arranged so that when some part of the generator is rotated, electric current moves through the wires.

The hand-crank generator you have used in previous labs is such a device.

Step 1: Disconnect the coil from the galvanometer and attach the leads of the hand-crank generator to the galvanometer. Slowly turn the handle until the meter responds.

a.   What does the galvanometer show?

_____

b.   What happens if you turn the handle the other way?

_____

The invention of the generator allowed the production of continually flowing electrical energy without the use of chemical batteries. Further developments led to the wide-scale distribution of electrical energy and the availability of household electricity.

Large-scale electrical distribution grids are powered by large-scale generators. These generators are commonly powered by steam turbines. The heat used to generate the steam is typically produced by burning coal or oil, or as a byproduct of controlled nuclear reactions.

Step 2: Use a generator to power a motor! Connect two hand-crank generators to each other. Crank the handle of one of the generators.

a.   What happens to the handle of the other generator? Would you say the other generator is acting as a motor?

_____

b.   Not all the energy you put into the generator turns into mechanical energy in the motor. What is your evidence of this?

_____

Summing Up

1.   Name each device described below

a.   Transforms electric energy into mechanical energy: _____

b.   Transforms chemical energy into electrical energy: _____

c.   Transforms mechanical energy into electrical energy: _____

2.   What happens to the energy lost between the generator and motor in Step 2 of Part B above?

_____

3.   A classmate suggests that a generator could be used to power a motor that could then be used to power the generator. What do you think about this proposal and why?

_____

_____

CONCEPTUAL PHYSICS | **Activity**

Pinhole Image

Purpose
To investigate the operation of a pinhole "lens" and compare it to the eye

Apparatus
3" × 5" card
straight pin
meterstick

Discussion
The image cast through a pinhole in a pinhole camera has the property of being in clear focus at any distance from the pinhole. That's because the tininess of the pinhole does not allow overlapping of light rays. (The tininess also doesn't allow the passage of much light, so pinhole images are normally dim as well.) When a pinhole is placed at the center of the pupil of your eye, the light that passes through the pinhole forms a focused image no matter where the object is located. Pinhole vision, although dim, is remarkably clear. In this activity, you will use a pinhole to see fine details more clearly.

Procedure
Step 1: Bring this printed page closer and closer to your eye until you cannot clearly focus on it any longer. Even though your pupil is small, your eye does not act like a true pinhole camera because it does not focus well on nearby objects.

Step 2: Poke a single pinhole into a card. Hold the card in front of your eye and read these instructions through the pinhole. Bright light on the print may be required. Bring the page closer and closer to your eye until it is a few centimeters away. You should be able to read the type clearly. Then quickly remove the card and see if you can still read the instructions without the benefit of the pinhole.

Summing Up
Enlist the help of people in your lab who are nearsighted and who are farsighted (if you're not one of them yourself).

1. A farsighted person without corrective lenses cannot see close-up objects clearly. Can a farsighted person without corrective lenses see close-up objects clearly through a pinhole?

   _____

2. A nearsighted person without corrective lenses cannot see faraway object clearly. Can a nearsighted person without corrective lenses see faraway objects clearly through a pinhole?

   _____

3. Why does a page of print appear sharper yet dimmer when seen through a pinhole?

   _____

Who would believe that so small a space could contain the image of all the universe? O mighty process! Here the figures, here the colors, here all the images of every part of the universe are contracted to a point. O what a point is so marvelous!

Leonardo da Vinci,
commenting on the *camera obscura*

CONCEPTUAL PHYSICS	Activity

Pinhole Camera

Purpose
To observe images formed by a simple convex lens and compare cameras with and without a lens

Apparatus
covered shoebox with a lid
25-mm converging lens
tracing paper
aluminum foil
masking tape

Discussion
The first camera used a pinhole opening to let light in. Because the hole was so small, light rays that entered could not overlap, which was precisely why a clear image was formed on the inner back wall of the camera. Because the opening was small, a long time was required to expose the film sufficiently. A lens allows more light to pass through and still focus the light onto the film. Cameras with lenses require much less time for exposure, and the pictures came to be called "snapshots."

Procedure
Step 1: Construct a camera as shown in Figure 1. It is a shoebox with a hole about an inch or so in diameter on one end, some tracing paper taped in the center to act as a screen, and an opening for viewing the screen on the other end. Tape some foil over the lens hole of the box. Poke a pinhole in the middle of the foil. Point the camera toward a brightly illuminated scene, such as the window during the daytime. Light enters the pinhole and falls on the tracing paper. Observe the image of the scene on the tracing paper.

Figure 1

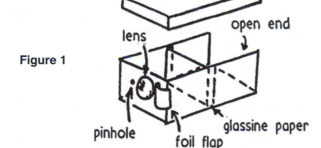

a.   Is the image on the screen upside down (inverted)?

_____

b.   Is the image on the screen reversed left to right?

_____

Step 2: Now remove the pinhole foil and tape a lens over the hole in the box. You now have a lens camera. Move it around and watch people or other scenery.

a. Is the image on the screen upside down (inverted)?

_____

b. Is the image on the screen reversed left to right?

_____

Step 3: Unlike lens cameras, pinhole cameras focus equally well on objects at practically all distances. Aim the camera lens at nearby objects and see if the lens focuses them.

a. Does the lens focus nearby objects as well as it does on distant ones?

_____

Step 4: Draw a ray diagram as follows. First, draw a ray for light that passes from the top of a distant object through a pinhole and onto a screen. Second, draw another ray for light that passes from the bottom of the object through the pinhole and onto the screen. Then sketch the image created on the screen by the pinhole.

Summing Up

1. Why is the image created by the pinhole dimmer than the one created by the lens?

_____

_____

_____

2. How is a pinhole camera similar to your eye? Do you think that the images formed on the retina of your eye are upside down? Your explanation might include a diagram.

_____

_____

_____

CONCEPTUAL PHYSICS	**Activity**

Image of the Sun

Purpose
To produce an image of the Sun on your classroom floor

Apparatus
index card
sharp point (pencil, pen, or equivalent for poking a hole in the index card)
small knife (penknife, X-acto knife, or equivalent for cutting a hole in the index card)
meterstick
access to direct sunlight

Discussion
A pinhole camera is a light-tight box with a small pinhole-sized opening in one end and a viewing screen at the opposite end. Light coming from an object that passes through the pinhole forms an image of the object. Why? Because rays of light coming from the object can't overlap and cause blurring. You can see in the sketch why the image is upside down.

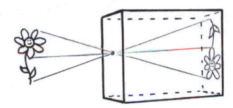

In this activity, your **pinhole camera** will simply be an index card with a hole punched in the middle by a pencil or pen. The object will be the Sun, bright enough to cast a clear image without a light-tight box.

Procedure
Step 1: Use the pencil or pen to poke a small hole in the center of the index card. The hole should be about 1 millimeter in diameter.

Step 2: Hold the card in bright sunlight about a meter or so above the floor. The card casts a shadow on the floor. In the middle of the card's shadow is a circular spot of light—an image of the Sun. If the Sun is low in the sky, the circular image is lengthened to an ellipse.

Step 3: Hold the card higher and the solar image becomes bigger.

Step 4: Repeat the procedure, but instead of a round hole poked in the card, carefully cut a diamond shape.

Does this affect the shape of the image?

_____

_____

_____

Step 5: Measure the diameter of the solar image and compare it to the distance between it and the pinhole. A convenient way to do this is to draw a circle 1 centimeter in diameter on a piece of paper. Place the paper on the floor where your solar image will fall. Then see how many centimeters high the pinhole needs to be for the solar image to fill the 1-cm circle.

How high is the pinhole above the paper? _____

Step 6: Determine and record how high the pinhole needs to be to produce a 2-cm wide image.

_____

Step 7: Determine and record how high the pinhole needs to be to produce a 10-cm wide image.

_____

Look at spots of light beneath sunlit trees. When the openings between leaves above are small compared with the distance to the ground below, images of the Sun are cast. What will be the shape of these images at the time of a partial solar eclipse?

| **CONCEPTUAL PHYSICS** | **Activity** |

Sunballs

Purpose
To estimate the diameter of the Sun

Apparatus
small piece of cardboard   1 dime   meterstick

Discussion

Take notice of the round spots of light on the shady ground beneath trees. These are sunballs—images of the Sun (featured in the Chapter 1 photo openers of your textbook) They are cast by openings between leaves in the trees that act as pinholes. The diameter of a sunball depends on its distance from the small opening that produces it. Large sunballs, several centimeters or so in diameter, are cast by openings that are relatively high above the ground, while small ones are produced by closer "pinholes." The interesting point is that the ratio of the diameter of the sunball to its distance from the pinhole is the same as the ratio of the Sun's diameter to its distance from the pinhole.

Because the Sun is approximately 150,000,000 km from the pinhole, careful measurement of this ratio tells us the diameter of the Sun. That's what this experiment is all about. Instead of finding sunballs under the canopy of trees, you'll make your own easier-to-measure sunballs.

Procedure

Step 1: Poke a small hole in a piece of cardboard with a pen or sharp pencil. Hold the cardboard in the sunlight and note the circular image that is cast on a convenient screen of any kind. This is an image of the Sun. Unless you're holding the card too close, note that the solar image size does not depend on the size of the hole in the cardboard (pinhole), but only on its distance from the pinhole to the screen. The greater the distance between the image and the cardboard, the larger the sunball.

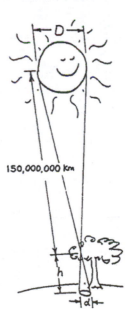

Step 2: Position the cardboard so the image exactly covers a dime, or something that can be accurately measured. Carefully measure the distance to the small hole in the cardboard. Record your measurements as a ratio:

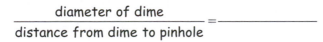

$$\frac{\text{diameter of dime}}{\text{distance from dime to pinhole}} = \underline{\hspace{3cm}}$$

Because this is the same ratio as the diameter of the Sun to its distance, then

$$\frac{\text{diameter of dime}}{\text{distance from dime to pinhole}} = \frac{\text{diameter of sun}}{\text{distance from sun to pinhole}}$$

Which means you can now calculate the diameter of the Sun!

Diameter of the Sun = _____

Summing Up

1. Will the sunball still be round if the pinhole is square shaped? Triangular shaped? (Experiment and see!)

_____

_____

_____

2. If the Sun is low in the sky so that the sunball is elliptical, should you measure the small or the long width of the ellipse for the sunball diameter in your calculation of the Sun's diameter? Why?

_____

_____

_____

3. If the Sun is partially eclipsed, what will be the shape of the sunball? (Note: The answer to this is shown on page 1 in the Eleventh Edition of *Conceptual Physics!*)

_____

_____

_____

4. Suppose you could fit 100 dimes, end to end, between your card with the pinhole and your dime-sized sunball. How many suns could fit between Earth and the Sun?

_____

_____

_____

CONCEPTUAL PHYSICS	Demonstration

Why the Sky Is Blue

Purpose
To see why the daytime sky is blue while the sunrise/sunset sky is red

Apparatus
a pair of resonant (sympathetic) tuning forks: matching tuning forks attached to wood sound boxes
tuning fork mallet
Laser Viewing Tank (or equivalent)
small, bright flashlight (e.g., LED Mini Maglite)
access to water
scattering agent
stirring rod
paper towel

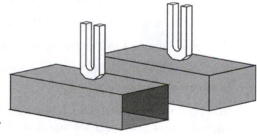

Discussion
In Chapter 27 in your textbook you read about why the sky is blue. In this activity, we extend the analogy of sound scattered by tuning forks to light scattered by *optical tuning forks,* the particles that make up the air.

BLUE SKY INGREDIENTS
The Earth's sky is made of air. Air is mostly molecular nitrogen (N_2) and molecular oxygen (O_2). Nitrogen and oxygen are neighbors on the periodic table; their molecules and have nearly the same size.

Still though, air is transparent and colorless. Air molecules are not blue, themselves.

At night, the sky appears transparent and colorless.

There is no "sky" on the Moon because the Moon has no atmosphere.

Our familiar blue sky requires both *air* and *sunshine*.

Procedure

PART A: RESONANCE AND SYMPATHETIC VIBRATION
Step 1: From the resonant pair of tuning forks, set one tuning fork aside.

Step 2: Using the tuning fork mallet, strike the other tuning fork and listen to the tone. A single frequency is produced by the tuning fork and amplified by the sound box it's attached to.

Step 3: Gently place your hand on the top of the tuning fork's tines to stop the vibration. Notice that the sound stops when you stop the tuning fork's vibrations.

Step 4: Now arrange the two tuning forks so that the open sides of their sound boxes face each other and are 10–20 cm apart.

Step 5: Strike one tuning fork. After a few seconds, stop the struck tuning fork's vibrations by gently placing your hand on top of its tines.

What do you hear *after* the struck tuning fork is stopped?

_____

_____

The struck tuning fork's vibrations create sound waves.

Those sound waves strike the second tuning fork. The frequency of the sound waves matches the frequency of the second tuning fork.

So the incoming sound waves set the second tuning fork into vibration.

The second tuning fork then sends out sound waves of its own.

When the struck tuning fork is stopped, the second tuning fork continues to vibrate and emit sound.

This process is a form of *scattering*: waves from one source excite another source, and the other source sends out waves as a result.

Step 6: Imagine a wide variety of tuning forks arranged in a certain region. And imagine a large number of identical tuning forks in a nearby region, as shown in Figure 1 below.

If all the tuning forks in the variety were struck, would the matching tuning forks in the other region be set into vibration? Explain.

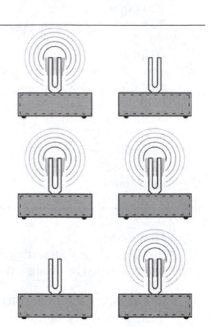

_____

_____

Variety of Tuning Forks   Identical Tuning Forks

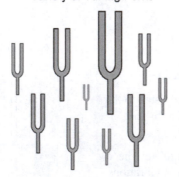

Figure 1

FROM THE SUN TO THE SKY
Step 7: Answer the following questions about the Sun's light and the Earth's atmosphere.
a.   Does the Sun emit one frequency of electromagnetic radiation or many frequencies? Describe.

_____

_____

Why the Sky Is Blue   236

b.   Do atmospheric molecules have a wide range of sizes or do they have a narrow range of sizes?

_____

c.   Therefore, would you expect atmospheric molecules to resonate with a wide variety of frequencies or with a narrow range of frequencies?

_____

d.   How does the lesson of the tuning forks apply to the scattering of light in the atmosphere?

_____

_____

_____

COLOR SENSITIVITY

The spectrum of visible light runs through the colors red, orange, yellow, green, blue, and violet. Our eyes are not equally sensitive to all colors of visible light.

They are most sensitive to colors near the center of the visible spectrum (yellow/green). This color, **chartreuse,** is used for most tennis balls, many roadside worker vests, and some fire trucks.

The sensitivity of our eyes diminishes for colors at the outer extremes of the spectrum: red and violet.

ATMOSPHERIC SCATTERING

The Sun emits all frequencies of the visible spectrum and frequencies beyond the visible spectrum, such as ultraviolet and infrared.

The molecular nitrogen and oxygen in the Earth's atmosphere resonate at frequencies that correspond to ultraviolet light. So the atmosphere scatters ultraviolet light better than light of any other frequency. But a substantial amount of the Sun's ultraviolet light is blocked by ozone high in the atmosphere. And ultraviolet light is not visible.

Violet light has a frequency near that of ultraviolet. So more violet light is scattered than light of any other color in the visible spectrum. Blue is scattered to a lesser degree. Green is scattered less than blue. Yellow is scattered less than green. Orange is scattered less than yellow. And red is scattered less than orange.

WHY THE SKY IS BLUE
Step 8: Solve the puzzle.
a.   Violet light is scattered more than blue light is. *Why is the sky not* **violet**?

_____

_____

_____

b.   Our eyes are more sensitive to green than they are to blue. *Why is the sky not* **green**?

_____

_____

_____

Why the Sky Is Blue

c. *Why is the sky **blue**?*

_____

_____

_____

PART B: SCATTERING IN THE TANK

Step 1: Fill the light beam viewing tank about three-quarters full with water.

Step 2: Gently stir enough scattering agent into the water so that it appears to have a slight visible haze to it. (This will be ***much*** more scattering agent than is used for other activities involving the tank.)

Step 3: Darken the room and use the flashlight to illuminate the tank from one end, as shown in Figure 2.

Step 4: Observe the tank from above. See position A shown in Figure 2. What color is observed in the light near the flashlight?

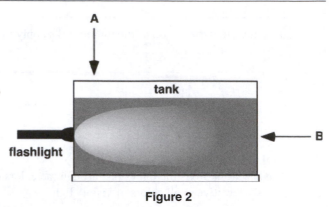

Figure 2

_____

Step 5: Observe the tank from the far end. See position B shown in Figure 2. What color is observed in the light that has passed through the full length of the tank?

_____

THE THICKNESS OF THE ATMOSPHERE

At midday, the Sun is at a high angle in the sky. Sunlight passes through a relatively thin layer of atmosphere before reaching the ground. In this thin layer, violet and blue are well scattered.

At sunset, sunlight passes through a thick layer of atmosphere before reaching the ground. At this greater thickness, the violet, blue, green, and yellow have been scattered.

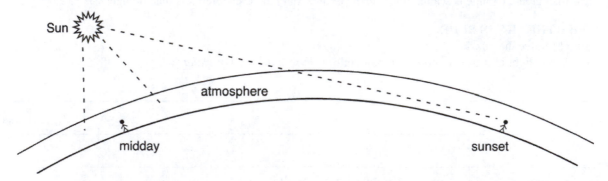

How is it that the same process can make the midday sky blue and the sunset sky red?

_____

_____

_____

CONCEPTUAL PHYSICS	**Activity**

Chapter 28: Reflection and Refraction   The Geometry of Plane Mirror Images

Mirror rorriM

Purpose
To investigate the minimum size mirror required for you to see a full image of yourself

Apparatus
large mirror, preferably full-length
ruler
masking tape

Discussion
Why do shoe stores and clothing shops have full-length mirrors? Must a
mirror be as tall and wide as you for you to see a complete image of yourself?

Procedure
Step 1: Stand about an arm's length in front of a vertical full-length mirror.
Reach out and place a small piece of masking tape on the image of the top of your head. Now stare at your toes.
Place the other piece of tape on the mirror where your toes are seen. Use a meter stick to measure the mirror's
distance from the top of your head to your toes, then measure your actual height. How does the distance between
the pieces of tape on the mirror compare with your height?

Step 2: Now stand about 3 meters from the mirror and repeat Step 1.
Stare at the top of your head and toes and have an assistant move the
tape so that the pieces of tape mark where your head and feet are seen.
Move further away or closer, and repeat. What do you discover?

Summing Up
1. Does the location of the tape depend on your distance from the mirror?

_____

_____

2. What is the shortest mirror you can use to see your entire image? Do you *believe* it?

_____

_____

Going Further

Try this one if a full-length mirror is not readily available *or* you are a disbeliever!
Hold a ruler next to your eye. Measure the height of a common pocket mirror.
Hold the mirror in front of you so that the image includes the ruler. How many
centimeters of the ruler appear in the image? How does this compare with the
height of the mirror?

_____

_____

_____

CONCEPTUAL PHYSICS	Activity

Trapping the Light Fantastic

Purpose
To investigate the behavior of light as it passes from one transparent material to another

Apparatus
Laser Viewing Tank (or equivalent)
opaque white tank insert/background
stirring rod
access to water
access to scattering agent (Mop & Glo, Pine-Sol, powdered milk, or equivalent)
laser pointer

Going Further
2-liter clear plastic bottle (soda pop, for example)

LASER LIGHT WARNING: TO AVOID INJURY, DO NOT EXPOSE EYES TO DIRECT LASER LIGHT. DO NOT AIM THE LASER AT PEOPLE OR ANIMALS.

Discussion
When light passes from one transparent material to another, it undergoes refraction. If the light crosses the boundary between the materials at an angle, the light changes direction. In this activity, you will see how the direction changes when light travels from air to water and when light travels from water to air.

Most optical devices—including glasses, contact lenses, cameras, microscopes, and telescopes—rely on lenses that refract light in the manner to be explored in Part A of this activity.

The refraction of light as explored in Part B of this activity is employed in fiber optics. Optical fibers are used in communication and medical technology.

Procedure

PART A: LIGHT BEAM TANK AND REFRACTION BASICS
If you shine a laser beam at a wall on the other side of the room, you will see the dot where the beam strikes the wall. The light is reflected diffusely and visible from nearly any place in the room. But the beam from the laser to the wall is not visible. Air is transparent. So the laser beam passes through the air without being scattered. If smoke or some other tiny particles are added to the air, the beam becomes visible. The ever-moving particles reflect the light of the beam in all directions.

Step 1: Fill the tank about halfway with water.

Step 2: Shine the laser into water, aim it through the length of the water, and record your observations.

_____

_____

Step 3: Add a small amount of scattering agent to the water and stir to mix it thoroughly.

Step 4: Shine the laser through the water again and record your observations. What—if anything—is different this time? Describe and explain.

_____

_____

_____

Step 5: Shine the laser beam so that it passes from air to water as shown in Figure 1 below.

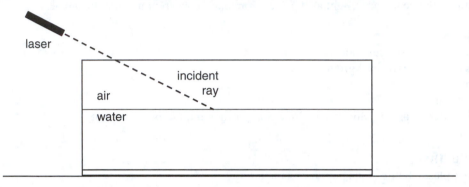

Figure 1

The incident ray is labeled in the diagram. Sketch and label the refracted ray (in the water) in the figure.

PART B: CRITICAL ANGLE AND TOTAL INTERNAL REFLECTION
Step 1: Move the tank a few centimeters over the edge of the table. Shine the laser beam so that it passes from water to air as shown in Figure 2. Sketch and label:

- the incident ray (in the water)

- the refracted ray (in the air above the water)

- the reflected ray (in the water)

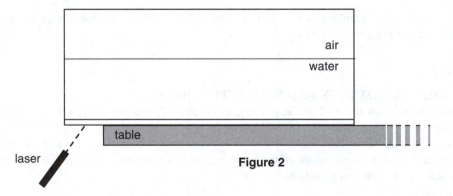

Figure 2

This arrangement is referred to as *subcritical.*

Step 2: Move and rotate the laser to change the angle of the incident ray as shown in Figure 3.

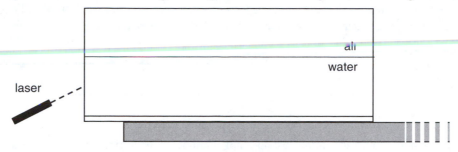

Figure 3

Sketch the result in the figure above and describe it in the space below.

_____

_____

Step 3: This arrangement is referred to as *supercritical.* The phenomenon it shows is *total internal reflection.* Why is this a good name? (*Hint:* Which ray from Figure 2 is no longer represented in Figure 3?)

_____

_____

_____

Step 4: Now that you have observed the *subcritical* and *supercritical* arrangements, find the *critical* arrangement. When the laser beam strikes the water-to-air boundary at the critical angle, the *refracted* ray travels along the surface of the water.

Going Further

Step 1: Obtain a clear, 2-liter plastic bottle. Cut off the top of the bottle and put a small, circular hole in the side of the bottle about 5 centimeters from the base of the bottle.

Step 2: Fill the container with tap water so that a steady stream flows outward from the hole. It's best to aim the stream into a sink.

Step 3: Shine the laser beam through the water to the hole as shown in Figure 4. What happens to the beam when it gets to the hole? Add a line to the figure and describe below.

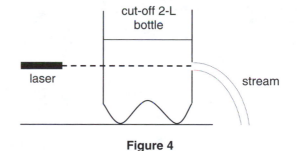

Figure 4

_____

_____

_____

Trapping the Light Fantastic

Summing Up

1. When light passes from air into water, it does not continue on its original path. Does it bend downward to travel deeper into the water or upward to travel at a shallower depth of water?

_____

_____

_____

2. When light passes from water into air, it does not continue on its original path. How does it bend?

_____

_____

_____

CONCEPTUAL PHYSICS	Activity

A Sweet Mirage

Purpose
To observe the behavior of light as it passes through a transparent material having a gradually changing index of refraction

Apparatus
Laser Viewing Tank (or equivalent)
access to hot water
access to scattering agent (Mop & Glo, Pine-Sol, powdered milk, or equivalent)
stirring rod
10 sugar cubes
laser pointer

LASER LIGHT WARNING: TO AVOID INJURY, DO NOT EXPOSE EYES TO DIRECT LASER LIGHT. DO NOT AIM THE LASER AT PEOPLE OR ANIMALS.

Discussion
As illustrated in Figures 28.19 through 28.22 in your textbook, mirages are the result of the gradual refraction that occurs when light passes through air in which its speed varies. Another way of saying this is that refraction occurs when the index of refraction varies. The refractive index of air depends on its temperature, and mirages are often seen where air near the ground is hot and air higher up is cooler.

In this activity, you will create an environment in which the index of refraction gradually changes. You will see how a beam of light behaves as it passes through such an environment.

Procedure
Step 1: Add hot water to the tank so that the tank is about two-thirds full.

Step 2: Add a small amount of scattering agent and stir the water to mix it.

Step 3: Add the sugar cubes so that they are fairly evenly spaced along the bottom of the tank as shown in Figure 1.

Step 4: Wait several minutes while the sugar dissolves. **It is important that you do not mix or otherwise disturb the water while the sugar dissolves.** Allow at least 10 minutes for the sugar to dissolve. Note that not all the sugar will dissolve in this time.

Step 5: Shine the laser beam horizontally through the water near the top of the water as shown in Figure 2. Draw the path of the beam as it passes through the water.

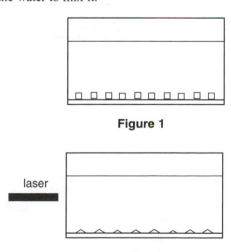

Figure 1

Figure 2

Step 6: Keep the laser beam horizontal and move it downward toward the bottom of the tank as shown in Figure 3. Draw the path of the beam when it no longer passes straight through the water.

Step 7: Upon approval from your instructor, stir the sugar-water solution to mix it thoroughly.

Step 8: Shine the laser beam horizontally through the water, high and low in the tank. Draw the path of the beam at the high and low positions shown in Figure 4.

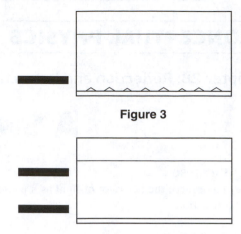

Figure 3

Figure 4

Summing Up

1. How did the concentration of sugar in the water vary from the top of the tank to the bottom after the cubes had been in the water for several minutes?

_____

_____

_____

2. Why did the beam of light behave the way it did in Step 6?

_____

_____

_____

3. Why did the beam behave the way it did in Step 8?

_____

_____

_____

CONCEPTUAL PHYSICS	**Demonstration**

Light Rules

Purpose
To use simple geometry and simple equipment to determine the wavelength of laser light

Apparatus
laser with a known wavelength
metal ruler with an etched millimeter scale
metersticks or tape measure

Discussion
If you view a meterstick face on, at right angles to your vision, millimeter marks appear simply as millimeter marks. But if you view the meterstick at a grazing angle, not only does the meterstick appear foreshortened, but the markings on the ruler appear "squashed" in your line of sight.

Likewise for a laser beam reflecting at a grazing angle from the surface of a tilted meterstick. When the stick is tilted as shown in Figure 1 (Step 1), the millimeter marks appear to be 1/10 of a millimeter to the light beam. One millimeter is seen as 0.10 mm. Likewise for the raised millimeter marks of a metal or plastic ruler. With this grating spacing and the viewing screen far away, we can use the raised ridges of the ruler as a diffraction grating and measure the wavelength of light!

The diffraction pattern produced by a transmission or reflection grating is an interference pattern that produces very distinct maxima (bright spots of constructive interference). The angle θ of the first-order maxima depend on the wavelength λ of the

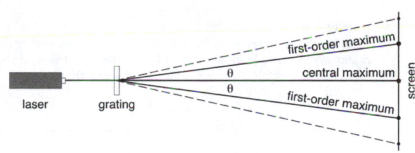

laser light and the line spacing d of the diffraction grating: $\lambda = d \sin\theta$. See Figure 29.18 and the footnote on page 517 of your textbook for a detailed discussion on the reason for the bright and dark zones.

Procedure
Step 1: Arrange a simple reflection. Shine a laser at a smooth (unmarked) section of the metal ruler, as shown in Figure 1. A reflected dot will appear on the wall or screen.

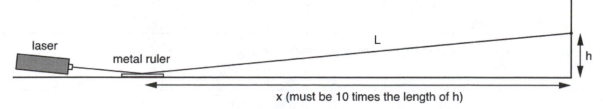

Figure 1. Specular Reflection

The laser beam's path has a slope of 1/10. That is, its height changes by 1 cm for every 10 cm that it travels forward.

The horizontal distance x must be 10 times the vertical distance h. Secure the laser once the configuration is correctly arranged. Record the distances x and h for your arrangement. Don't forget to include units! The ruler-to-screen distance L will be determined later.

$x =$ _____ $h =$ _____

Step 2: Move the ruler so that the laser beam strikes the etched millimeter marks. A diffraction pattern will now appear as shown in Figure 2.

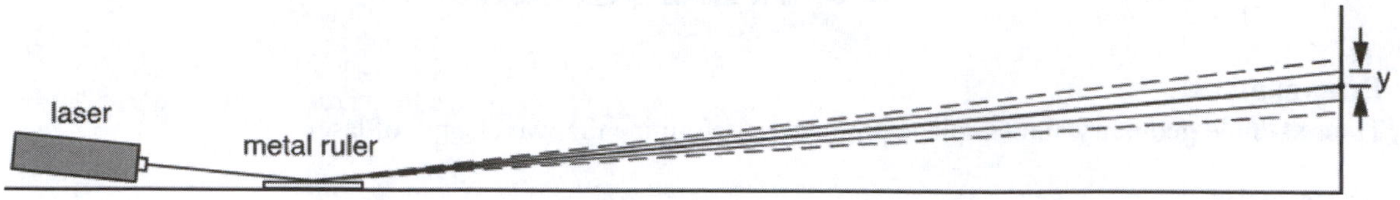

Figure 2. Diffraction

Step 3: Measure the distance y to the first-order maxima. The pattern of dots will be somewhat asymmetric due to the oblique angles used in the arrangement. To get a good value for the first-order maxima distance, do the following.

Measure the distance from one first-order maximum to the other. See Figure 3. Take this distance to be $2y$. Find y by dividing that distance by 2.

$2y =$ _____ $y =$ _____

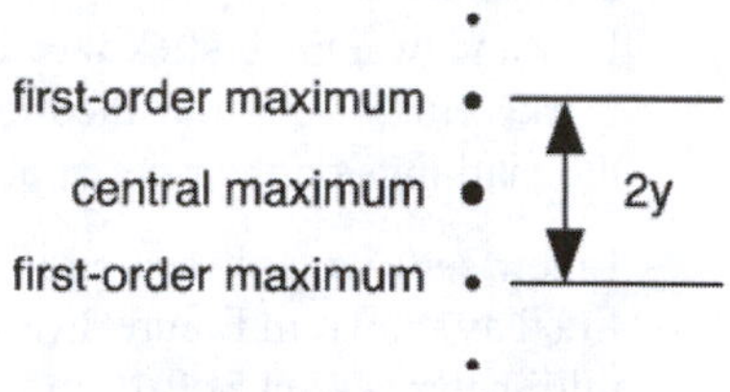

Step 4: Record the wavelength of the laser in nanometers and in meters (use scientific notation).

$\lambda =$ _____ nm = _____ m

Figure 3. Maxima

Summing Up

1. Examine Figure 1. Since x is 10 times the length of h, can you see that the grating-to-screen distance L is **essentially** the same as the distance x? Record the value below. If you are concerned that this is an unwarranted simplification, determine the value of L from the values for x and h using the Pythagorean Theorem.

    $L =$ _____

2. Calculate the wavelength of the laser using $\lambda = d \sin\theta$.

    a. The value of $\sin\theta$ in the geometry of our skinny triangle can be taken to y/L.

    $\sin\theta = y/L =$ _____

    b. Multiply this result by d. Note that d is the foreshortened space between our etched millimeter marks. It's $1/10^{th}$ of 1 mm, which is 1 µm, or 1×10^{-6} m.

    $\lambda = d\sin\theta = dy/L =$ _____

    c. Express the wavelength to the nearest nanometer (nm).

    $\lambda =$ _____

3. Calculate the percent difference between this value and the accepted wavelength of the laser.

    _____

    _____

CONCEPTUAL PHYSICS	Activity

Diffraction in Action

Purpose
To observe the wave nature of light as it passes through a single, narrow slit and as it passes through many narrow slits

Apparatus
point source of incandescent light (*incandescent* Mini Maglite or equivalent)
2 pencils (or equivalent)
diffraction grating
compact fluorescent light (illuminated)

Going Further
white LED light (LED Mini Maglite, for example)

Discussion
We are accustomed to light traveling in straight-line paths. We don't expect light to bend around corners or to spread out after passing through small holes. Yet it is capable of doing both! In this activity, you will see the effect of light spreading out after passing through a narrow opening. And you will see that different colors spread out by different amounts.

Procedure
Step 1: Prepare the point source. For example, remove the lens housing from the body of the Mini Maglite by twisting it counterclockwise until it comes off. When you are done, the point source (bulb) should be glowing brightly. You can use the lens housing as a stand for the Mini Maglite. See Figure 1.

When you look at the point source, you will see a **dandelion** of light around it. Where is the dandelion, actually? Is it really around the bulb of the point source, or is it in your eye? While observing the dandelion, rotate the flashlight then rotate your head. Where is the dandelion and how do you know?

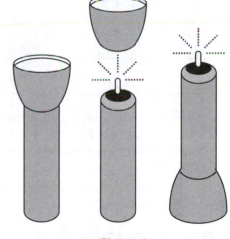

Figure 1

_____

_____

_____

_____

Step 2: Hold the two pencils so they are side-by-side and touching each other. Look through the tiny space between them toward the point source. Record your observations using words and pictures.

_____

_____

_____

Step 3: Make the gap between the pencils smaller by squeezing the pencils against one another. How—if at all—does this change what you see? (Does the pattern get narrower, wider, or remain the same?)

_____

_____

The pattern you see is the result of light wave interference. Light passing through a narrow opening spreads out. As the light spreads out, an interference pattern results in the bright and dark bands or "fringes." This phenomenon is called *single-slit diffraction.*

Step 4: Set the single-slit materials aside and view the point source by looking through the diffraction grating (film slide). The diffraction grating has thousands of slits packed closely together. The phenomenon that results is called *multiple-slit diffraction.*

a.   Record your observations in words and pictures.

_____

_____

_____

_____

_____

The point source emits light in all the colors of the spectrum. Viewed on its own, light from the bulb appears white. The colors of the point source light do not diffract equally. Some are diffracted less and remain closer to the point source when viewed through the grating. Some colors are diffracted more and appear farther from the point source.

b.   List the colors of the spectrum (blue, green, orange, red, violet, yellow) in order, from least diffracted to most diffracted.

LEAST DIFFRACTED   MOST DIFFRACTED

_____   _____   _____   _____   _____   _____

Step 5: Turn off the point source and set it aside. Turn on the compact fluorescent light and observe it through the diffraction grating. Record your observations in words and pictures.

_____

_____

_____

_____

_____

_____

Going Further

Obtain a white LED and observe its light through the diffraction grating. Is the light from the LED more like the light from the point source (flashlight) or the compact fluorescent light? Justify your answer.

_____

_____

_____

Summing Up

1. Does the amount of diffraction increase or decrease as the wavelength of light increases? (*Hint:* Among the colors in the spectrum, violet light has the shortest wavelength and red light has the longest wavelength.)

_____

_____

2. If you weren't sure whether a light source was incandescent (like the point source) or fluorescent, could you use a diffraction grating to help make the determination? Explain.

_____

_____

Diffraction in Action

When you make the finding yourself—even if you're the last person on Earth to see the light—you'll never forget it.

Carl Sagan

Name _____ Section _____ Date _____

CONCEPTUAL PHYSICS

Activity

Laser Tree

Purpose
To determine the role of wavelength and line spacing in the geometry of interference patterns produced by diffraction gratings

Apparatus
Laser Viewing Tank (or equivalent)
access to water
access to scattering agent (Mop & Glo, Pine-Sol, powdered milk, or equivalent)
stirring rod
long-wavelength laser pointer (red)
short-wavelength laser pointer (green, blue, or violet)
low-density diffraction grating (~500 ± 100 lines/millimeter)
high-density diffraction grating (~1000 ± 100 lines/millimeter)
access to adhesive tape

Going Further
diffraction ("rainbow") glasses
airborne scattering agent ("fog in a can," professional haze, or equivalent)

LASER LIGHT WARNING: TO AVOID INJURY, DO NOT EXPOSE EYES TO DIRECT LASER LIGHT. DO NOT AIM THE LASER AT PEOPLE OR ANIMALS.

Discussion
When light passes through a pair of thin slits, an interference pattern reveals the wave nature of light. A diffraction grating is a film with thousands of slits. Light waves passing through a diffraction grating also produce an interference pattern. The geometry of the pattern depends on the spacing of the slits and the wavelength of the light.

Procedure
Step 1: Fill the tank with water to within about an inch of its full capacity.

Step 2: Add a small amount of scattering agent to the water and stir to mix it thoroughly.

Step 3: Tape the **high-density** diffraction grating to one end of the tank as shown in Figure 1. The grating lines must be horizontal so that light passing through them spreads out vertically.

(Note: The slits are very close together on the high-density grating; more lines per millimeter means less space between the lines.)

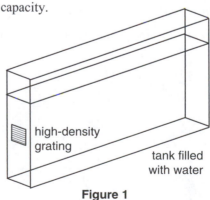

high-density grating

tank filled with water

Figure 1

Laser Tree

Step 4: Shine the short-wavelength (green) laser through the grating and observe the pattern in the tank. Sketch the pattern in Figure 2.

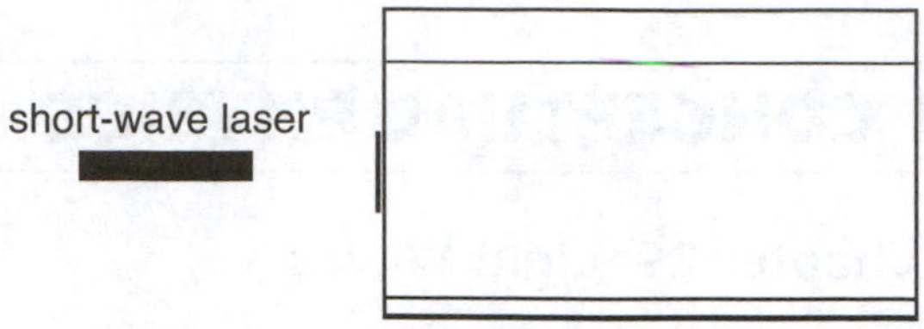

Step 5: Shine the long-wavelength (red) laser through the grating and observe the resulting pattern. Sketch the pattern in Figure 3.

How is the geometry of the long-wavelength diffraction pattern different from that of the short-wavelength pattern?

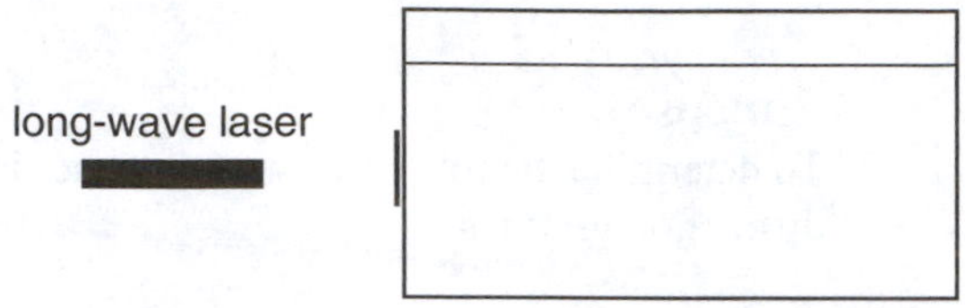

short-wave laser

Figure 2

_____

_____

long-wave laser

_____

Figure 3

_____

Step 6: Replace the high-density grating with the *low-density* grating. The space between the slits is greater on the low-density grating.

Step 7: Shine the short-wavelength (green, blue, or violet) laser through the grating and observe the pattern in the tank. Sketch the pattern in Figure 4.

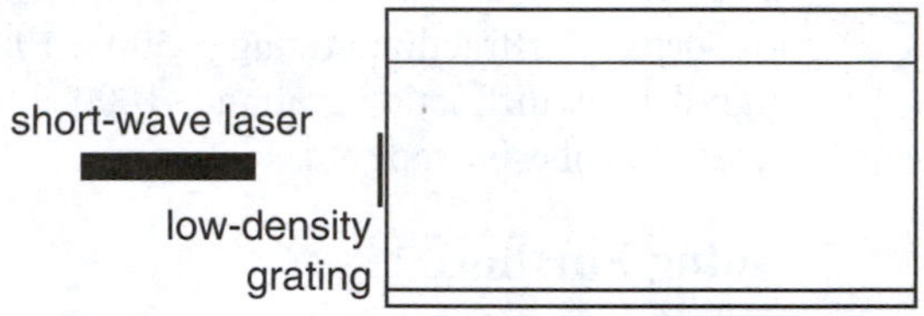

How is the geometry of the high-density grating diffraction pattern different from that of the low-density grating pattern?

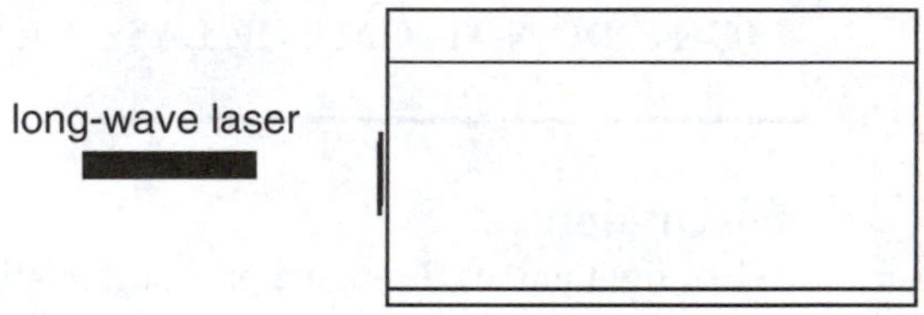

short-wave laser

low-density grating

Figure 4

_____

_____

long-wave laser

_____

_____

Figure 5

Step 8: Shine the long-wavelength (red) laser through the grating and observe the resulting pattern. Sketch the pattern in Figure 5.

Going Further
Step 1: With room lights dimmed or out, create a "cloud" of airborne scattering agent.

Step 2: Shine the green laser through the diffraction ("rainbow") glasses and into the cloud of scattering agent.

What is the result, and how is it different from the pattern seen in the previous steps? Describe and sketch the pattern in the space below.

_____

_____

_____

_____

Step 3: While the laser is shining through the glasses, rotate the glasses clockwise or counterclockwise.

What effect—if any—does rotating the glasses have on the pattern?

_____

_____

_____

_____

Summing Up

1. Which combination of grating and light produces the widest pattern (that is, the beams are most spread out)?

_____

_____

2. Which combination of grating and light produces the narrowest pattern?

_____

_____

3. As the wavelength of light increases (for example, from green to red), what happens to the width of the pattern?

_____

_____

4. As the line spacing on the grating increases (for example, from 1,000 lines per millimeter to 500 lines per millimeter), what happens to the width of the pattern?

_____

_____

The role of the school is not to make ideas safe for students,
but to make students safe for ideas.

| **CONCEPTUAL PHYSICS** | **Demonstration** |

Pole-Arizer

Purpose
To see the production of polarized and unpolarized mechanical waves and to see how unpolarized waves become polarized

Apparatus
4 long support rods (approximately 1 meter long)
4 short crossbars (approximately 30 centimeters long)
8 right-angle clamps
1 25-ft (8-m), coiled telephone cord (or equivalent)
3 student volunteers (in addition to the instructor)

Discussion
Waves can undergo a number of processes: reflection, refraction, interference, and diffraction. Polarization is another wave phenomenon, the only one that occurs only in transverse waves. Applications of light wave polarization include 3-D movie production and the optics of sunglasses. Before exploring those applications, it's a good idea to understand the process of polarization. Because light waves cannot be observed, we will observe mechanical waves in a phone cord.

Procedure
Step 1: Prepare *two* devices like those shown in Figure 1.

a. Use two right-angle clamps to attach two crossbars to a support rod.

b. Use two more right-angle clamps to attach the two crossbars to a second support rod.

c. Repeat to make a second device.

Step 2: Ask a student volunteer to hold one end of the phone cord at chest height. The cord should be held still throughout the demonstration.

Step 3: Shake the cord vertically. Adjust the tension in the cord by changing the distance to the volunteer holder. A pulse should take approximately 1 second to get from the shaker to the holder. If the tension is too great, it's difficult to create waves with adequate amplitude. If the tension is too low, the cord will droop and hit the ground.

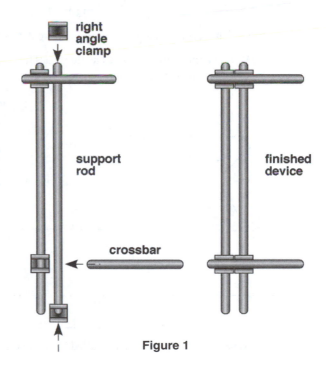

Figure 1

a. When the cord is shaken vertically, the resulting wave is ***polarized***. Observers to the side can see the crests and troughs of the cord, but an observer looking down from above would see no waves.

b. When the cord is shaken in a circle, the resulting wave is ***unpolarized***. Observers to the side and observers looking down from above would all see the crests and troughs.

c. What kind of wave—polarized or unpolarized—is made when the cord is shaken *horizontally?* Justify your answer.

Step 4: Select a second volunteer to hold one of the devices. Feed the phone cord through the gap and place the volunteer so they will be halfway between you and the holder. Ask the volunteer to use the handles to hold the structure vertically. See Figure 2. Position the device so that the stretched cord passes through the gap without touching very much. See Figure 3.

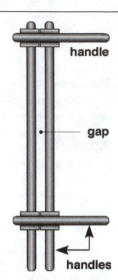

Figure 2

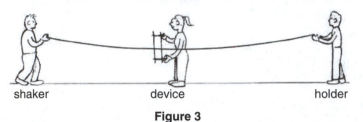

Figure 3

Step 5: The shaker shakes the cord vertically. The device should have no effect on the waves as they pass through to the cord holder. See Figure 4.

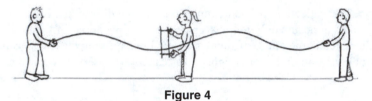

Figure 4

Step 6: The shaker shakes the cord in a circular pattern to create unpolarized waves.

a. What happens to the waves as they pass through the device? Complete Figure 5 and describe the result in words in the space below.

Figure 5

b. What should the device be called? See the title of the demonstration for a hint.

Because the structure is made of poles, you can see how the name came about! For purposes of simplicity, we will refer to the device as a *polarizer* for the remainder of the demonstration.

Step 7: Change the orientation of the polarizer so that it is horizontal. Make sure the polarizer volunteer holds the device in a way that her arms or hands don't interfere with the progress of incoming waves. What effect does the device have on unpolarized waves now?

_____

_____

Step 8: Select a *third* volunteer to hold the other polarizer. Again, feed the cord through the polarizer and arrange the volunteers so that the cord distance is divided into thirds. Have both polarizers oriented vertically to begin with. See Figure 6.

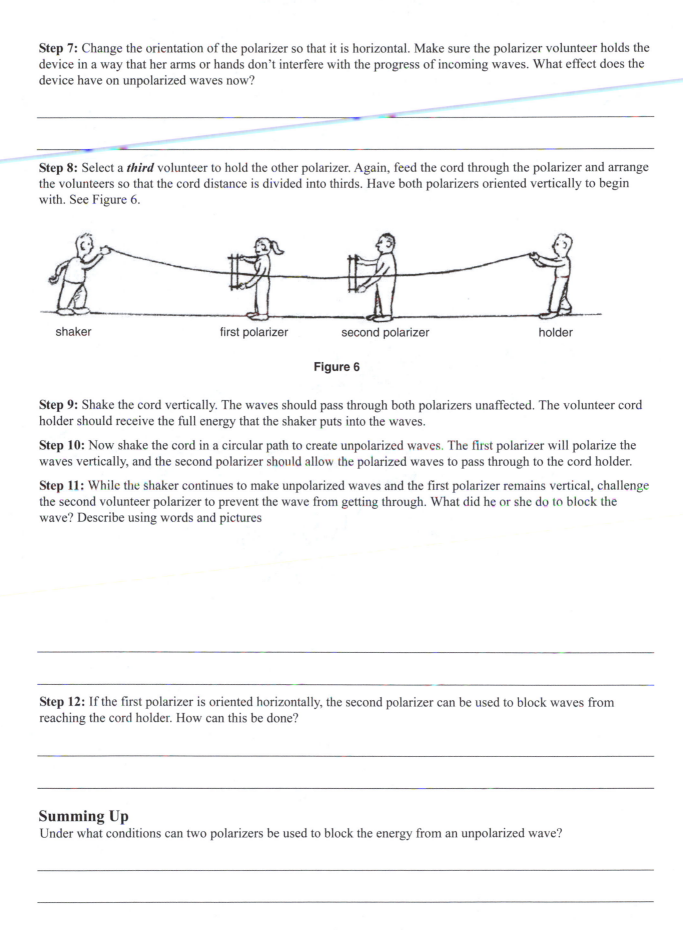

shaker　　　　　　first polarizer　　　　second polarizer　　　　　holder

Figure 6

Step 9: Shake the cord vertically. The waves should pass through both polarizers unaffected. The volunteer cord holder should receive the full energy that the shaker puts into the waves.

Step 10: Now shake the cord in a circular path to create unpolarized waves. The first polarizer will polarize the waves vertically, and the second polarizer should allow the polarized waves to pass through to the cord holder.

Step 11: While the shaker continues to make unpolarized waves and the first polarizer remains vertical, challenge the second volunteer polarizer to prevent the wave from getting through. What did he or she do to block the wave? Describe using words and pictures

_____

_____

Step 12: If the first polarizer is oriented horizontally, the second polarizer can be used to block waves from reaching the cord holder. How can this be done?

_____

_____

Summing Up
Under what conditions can two polarizers be used to block the energy from an unpolarized wave?

_____

_____

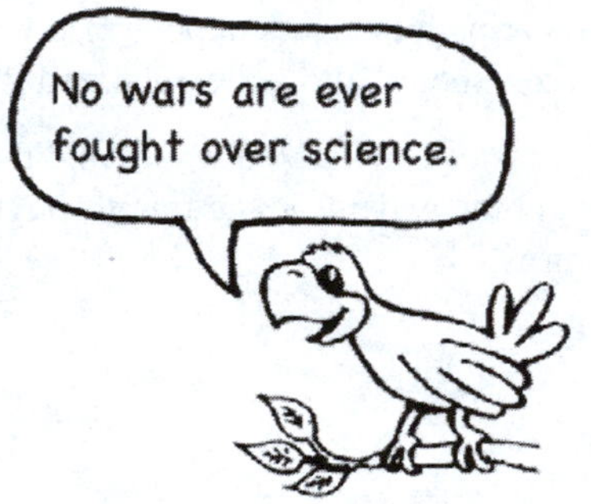
No wars are ever fought over science.

CONCEPTUAL PHYSICS	Demonstration

Chapter 29: Light Waves

Blackout

Purpose

In this demonstration, you will see the effects of the polarization of light waves. Specifically, unpolarized light will be passed through a polarizer. The polarized light will be passed through a second polarizer. Some other objects will be placed between the two polarizers to produce surprising results.

Apparatus

bright light source such as an overhead projector or slide projector (do **not** use an LCD projector)
large sheet of polarizing film (large polarizers)
large polarizer backed with a diffuser (thin sheet of white paper or equivalent)
transparent plastic objects such as protractors, forks, etc.
small polarizing filter

Going Further

apparatus from previous demonstration, "Pole-Arizer"

Discussion

In "Pole-Arizer," you saw how an unpolarized, mechanical wave could be polarized. And you saw the result of using multiple polarizers. So you know that polarized waves propagate with a preferred axis of oscillation. Light waves are electromagnetic waves in which electric and magnetic fields oscillate perpendicular to one another as the wave propagates perpendicular to those oscillations. Figure 1.a. is a representation of a single, polarized ray of light. Figure 1.b. shows the electric field oscillation, alone. Figure 1.c. shows a head-on view of the electric field oscillation.

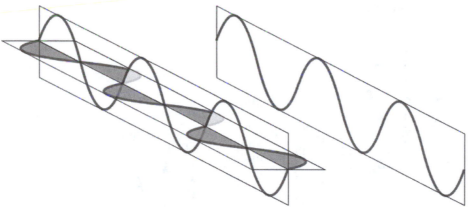

Figure 1.a. The electric field oscillates vertically while the magnetic field oscillates horizontally

Figure 1.b. The electric field, alone

Figure 1.c. Head-on electric field

Blackout

Most natural light is unpolarized, meaning electric field oscillations occur in many orientations as the wave propagates forward, as shown in Figure 2.a. A polarized beam is one in which a single axis of electric field oscillation is preferred, as shown in Figure 2.b.

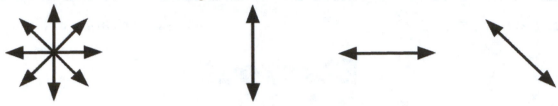

Figure 2.a. Unpolarized light with many directions of electric field oscillation

Figure 2.b. Three examples of polarized light, where one direction of electric field oscillation is preferred

In this demonstration, you will observe the effects of polarizing light waves. You will observe cross-polarization and try to solve a polarization puzzle. You will also see optical activity in transparent plastic.

Procedure

Step 1: Shine the light beam at the diffuser-backed large polarizer as shown in Figure 3. The diffuser allows observers to see the demonstration from a wide variety of angles. The polarizer allows one axis of electric field oscillation through.

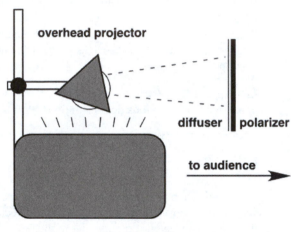

Figure 3

An *ideal* polarizer would block approximately 50% of unpolarized light incident upon it. See Figure 4 for an illustration. Note that each diagonal oscillation has both vertical and horizontal components. So some part of a diagonal oscillation can pass through a vertical polarizer, and some part of a diagonal oscillation can pass through a horizontal polarizer.

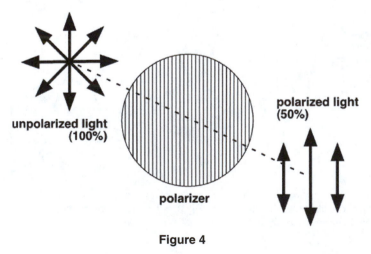

Figure 4

Observe that there is nothing peculiar about the polarized light. It may have a slight green-gray color to it. But for the most part, it appears to be nothing more than a diminished brightness of "regular," unpolarized light.

Step 2: Add a *second* polarizer (with no diffuser) in line with the first polarizer so that the optical axis (direction of polarization) matches that of the first.

Which is the better description of light passing through polarizers when their optical axes are parallel?
____Most of the light that passes through the first polarizer also passes through the second polarizer.
____Very little of the light that passes through the first polarizer also passes through the second polarizer.

Step 3: Rotate the second polarizer until its optical axis is *perpendicular* to that of the first.

Which is the better description of light passing through polarizers when their optical axes are perpendicular?
____Most of the light that passes through the first polarizer also passes through the second polarizer.
____Very little of the light that passes through the first polarizer also passes through the second polarizer.

Step 4: With the polarizers crossed, place a small polarizing filter *between* them. Orient the small filter so that its optical axis is at a 45° angle to both large polarizers. What do you see?

_____

_____

_____

_____

Consider the sequence illustrated in Figure 5 as you ponder why this happens.

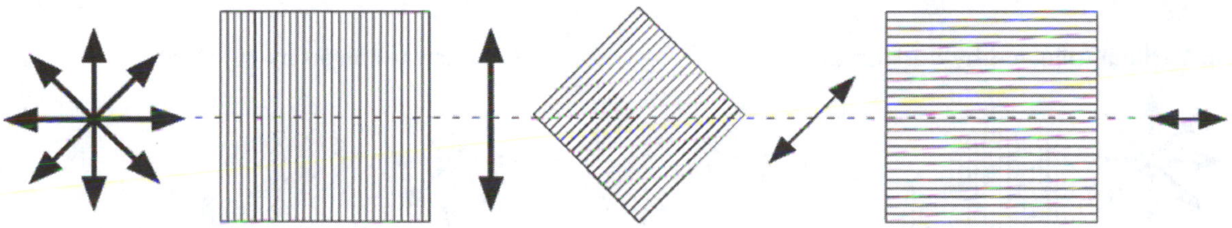

Figure 5

a. Unpolarized light is polarized when passing through the vertical filter.

b. Some of the vertically polarized light passes through the 45°-angle polarizer.

c. Some of the 45°-polarized light can pass through the horizontal

Step 5: With the polarizers crossed (optical axes perpendicular to one another), place a transparent plastic object *between* them. (The object might be a protractor, ruler, fork, or a similar thin, solid object.) What do you see?

_____

_____

_____

Blackout

Summing Up

1. Complete the illustrations of electric field oscillation below to show what happens in each polarizer configuration. Where two polarizers are used, show the oscillation *between* the filters as well as the final output.

a. One vertical filter (result is shown)

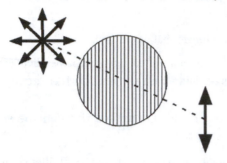

b. One horizontal filter

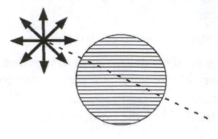

c. Vertical followed by another vertical

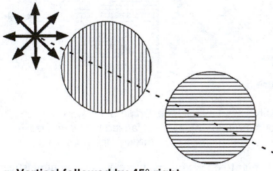

d. Horizontal followed by horizontal

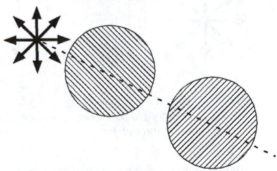

e. Vertical followed by horizontal

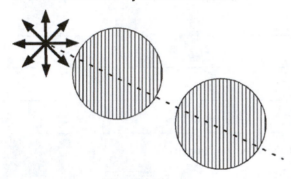

f. 45° left followed by 45° right

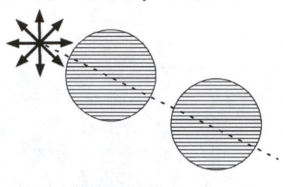

g. Vertical followed by 45° right

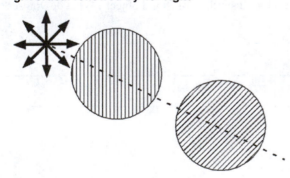

h. 45° right followed by horizontal

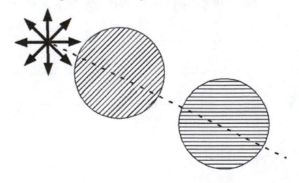

2. How might certain colors of light make it past two crossed polarizers as they did in Step 5? Describe in words and pictures.

_____

_____

_____

_____

_____

_____

_____

_____

Going Further

1. Do some research on "optical activity" to find out why the plastic objects allow polarized light to get through a perpendicular second polarizer. Record your findings on a separate sheet.

2. Can you use materials from "Pole-Arizer" to simulate the demonstration in Step 4? That is, can you use a polarizer to get energy from vertically polarized waves through a horizontal polarizer? If not, why not; if so, how?

_____

_____

_____

_____

Science is organized common sense where many a beautiful theory was killed by an ugly fact.

Thomas Huxley

CONCEPTUAL PHYSICS　　　　　　**Experiment**

Chapter 30: Light Emission　　　　　　　　　　　　　　　Spectroscopy

Bright Lights

Purpose
To examine light emitted by various elements when heated and identify a component in an unknown salt by examining the light emitted by the heated salt

Apparatus

Equipment
Bunsen burner
heatproof glove or potholder
metal spatulas or flame test wires (nichrome)
diffraction gratings
spectroscope (commercial or homemade)
ring stand with clamp large enough to fit around
spectroscope barrel
colored pencils
discharge tubes containing various gases

Chemicals
0.1 M HCl solution
sodium chloride powder
potassium chloride powder
calcium chloride powder
strontium chloride powder
barium chloride powder
cupric chloride powder

Discussion

When materials are heated, their elements emit light. The color of the light is characteristic of the type of elements in the heated material. Strontium, for example, emits predominantly red light, and copper emits predominantly green light. An element emits light when its electrons make a transition from a higher energy level to a lower energy level. Every element has its own characteristic pattern of energy levels and, therefore, emits its own characteristic pattern of light frequencies (colors) when heated.

It is interesting to look at the light emitted by elements through either a diffraction grating or a prism. Rather than producing all colors, the elements produce a spectrum showing only particular colors (frequencies). When the emitted light passes first through a thin slit, and then through the grating or prism, the different colors appear as a series of vertical lines, as shown in Figure 1. (The vertical lines are images of the original slit.) Each vertical line corresponds to a particular energy transition for an electron in an atom of the heated element. The pattern of lines, called an **emission spectrum,** is characteristic of the element. It is often used as an identifying feature—much like a fingerprint. Astronomers, for instance, can tell the elemental composition of stars by examining their emission spectra.

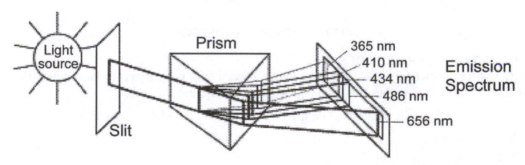

Figure 1. When heated, a gaseous element produces a discontinuous emission spectrum.

Procedure

PART A: FLAME TESTS

Step 1: Using either a heatproof glove or a potholder, hold the tip of a metal spatula (or nichrome wire, if available) in a Bunsen burner flame until the spatula tip is red hot, and then dip it in a 0.1 M HCl solution. Repeat this cleaning process several times until you no longer see color coming from the metal when it is heated.

Step 2: Obtain small amounts of the metal salts to be tested, and label each sample. Dip the spatula tip first into the HCl solution and then into one of the salts, so that the tip becomes coated with the powder. Then put the tip into the flame and observe the color. Record your observations on the report sheet.

Step 3: Rinse the spatula in water, and then clean as described in Step 1.

Step 4: Repeat this procedure for all the salts, being sure to clean the spatula each time.

PART B: FLAME TESTS USING A DIFFRACTION GRATING

Step 1: Look at the lights in your classroom through a diffraction grating. Note how the diffraction grating separates the white light into a rainbow of all colors.

Step 2: After your instructor darkens the room lights, repeat the procedure of Part A. This time, however, observe the flame through your diffraction grating. Note how the flame produced by the salt only produces a select number of colors.

Step 3: To help you distinguish which colors the salts are emitting, use a spectroscope. You can use either a commercial spectroscope or a homemade one like the one shown in Figure 2.

Step 4: On your report sheet, sketch the predominant lines you observe for each salt, using colored pencils. You will see lines both to the left and to the right of the slit. Sketch only the lines to the right. (Note: Some salts will also show regions of continuous color.)

Step 5: Obtain an unknown metal salt from your instructor and record its number.

Step 6: Observe and sketch the line spectrum for your unknown salt, using colored pencils.

Step 7: Identify your unknown salt, based on its spectrum.

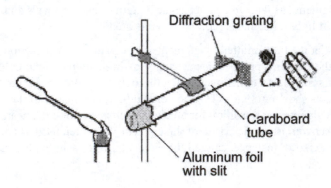

Figure 2. A homemade spectroscope.

PART C: GAS DISCHARGE TUBES

Step 1: Through a diffraction grating, observe the light emitted from various gas discharge tubes. (You need not pass the light through a slit because the discharge tubes themselves are narrow.)

Step 2: Using colored pencils, sketch the line spectra for all samples available, especially hydrogen, oxygen, and water vapor.

Bright Lights Report Sheet

PART A: FLAME TEST

Compound	NaCl	KCl	CaCl$_2$	SrCl$_2$	BaCl$_2$	CuCl$_2$
Color						

PART B: FLAME TESTS USING A SPECTROSCOPE

NaCl	Sketch of line spectrum
	V I B G Y O R

SrCl$_2$	Sketch of line spectrum
	V I B G Y O R

KCl	Sketch of line spectrum
	V I B G Y O R

BaCl$_2$	Sketch of line spectrum
	V I B G Y O R

CaCl$_2$	Sketch of line spectrum
	V I B G Y O R

CuCl$_2$	Sketch of line spectrum
	V I B G Y O R

Unknown number _____

Unknown	Sketch of line spectrum
	V I B G Y O R

Identity of unknown: _____

PART C: DISCHARGE TUBES

Substance	Sketch of line spectrum
	V I B G Y O R

Substance	Sketch of line spectrum
	V I B G Y O R

Substance	Sketch of line spectrum
	V I B G Y O R

Substance	Sketch of line spectrum
	V I B G Y O R

Substance	Sketch of line spectrum
	V I B G Y O R

Substance	Sketch of line spectrum
	V I B G Y O R

Substance	Sketch of line spectrum
	V I B G Y O R

Substance	Sketch of line spectrum
	V I B G Y O R

Summing Up

1. When the spatula was initially being cleaned in the flame, it may have given off yellow light. If this happened, what residue was probably on the spatula before it was cleaned?

_____

_____

2. What produces the colors of fireworks?

_____

_____

3. Is the gas in a blue "neon lamp" actually neon? Explain.

_____

_____

4. Does the line spectrum of water vapor bear any resemblance to the line spectra of hydrogen and oxygen? Why or why not?

_____

Name _____   Section _____   Date _____

CONCEPTUAL PHYSICS	Activity

Get a Half-Life

Purpose
To simulate radioactive decay half-life

Apparatus
25 small color-marked cubes per group (one side red, two sides blue, three sides blank)
spray-painted sugar cubes or multifaceted dice can be used

Discussion
The rate of decay for a radioactive isotope is measured in terms of half-life—the time for one-half of a radioactive quantity to decay. A graphical illustration of half-life is shown in Figure 33.16 in your textbook. Each radioactive isotope has its own characteristic half-life (Table 1). For example, the naturally occurring isotope of uranium, uranium-238, decays into thorium-234 with a half-life of 4.5 billion years. This means that only half of an original amount of uranium-238 remains after this time. After another 4.5 billion years, half of this decays leaving only one-fourth of the original amount remaining. Compare this with the decay of polonium-214, which has a half-life of 0.00016 seconds. With such a short half-life, any sample of polonium-214 will quickly disintegrate.

Table 1

Isotope	Half-life
Uranium-238	4,500,000,000 years
Plutonium-239	24,400 years
Carbon-14	5,730 years
Lead-210	20.4 years
Bismuth-210	5.0 days
Polonium-214	0.00016 seconds

The half-life of an isotope can be calculated by the amount of radiation coming from a known quantity. In general, the shorter the half-life of a substance, the faster it decays, and the more radioactivity per amount is detected.

In this activity, you will investigate three hypothetical substances, each represented by a color on the face of a cube. The first substance, represented by a given color, is marked on only one side of the cube. The second substance, represented by a second color, is marked on two sides of the cube, and the third substance, represented by a third color (or lack thereof), is marked on the remaining three sides. Rolling a large number of these identically painted cubes simulates the process of decay for these substances. As a substance's color turns face up, it is considered to have decayed and is removed from the pile. This process is repeated until all of the cubes have been removed. Since the color of the first substance is only on one side, this substance will decay the slowest (because its color will fall face up least frequently and it will stay in the game longer). The second substance, marked on two sides, will decay faster requiring fewer rolls before all the cubes are removed. The third substance, marked on three sides, will decay the fastest. After tabulating and graphing the numbers of cubes that decay in each roll for these simulated substances, you will be able to determine their half-lives.

Procedure

Step 1: Shake the cubes in a container and roll them onto a flat surface.

Step 2: Count the one-side color faces that are up and record this number under "Removed" in the data table.

Step 3: Remove the one-side color cubes in a pile off to the side.

Step 4: Gather the remaining cubes back into the container and roll them again.

Step 5: Repeat Steps 2–4 until all cubes have been counted, tabulated, and set aside.

Step 6: Repeat Steps 1–5 removing cubes that show the two-side color faces up.

Step 7: Repeat Steps 1–5 removing cubes that show the three-side color faces up.

Data Table

Throw	First Substance (One-side color)		Second Substance (Two-side color)		Third Substance (Three-side color)	
	Removed	Remaining	Removed	Remaining	Removed	Remaining
Initial Count						
1						
2						
3						
4						
5						
6						
7						
8						
9						
10						
11						
12						
13						
14						
15						
16						
17						
18						
19						
20						
21						
22						
23						
24						
25						

Step 8: Plot the number of cubes remaining versus the number of throws for each substance on the following graph. Use a different color or line pattern to graph the results for each substance. For each substance, draw a single smooth line or curve that approximately connects all points. ***Do not connect the dots!*** Indicate your color or line pattern code below the graph.

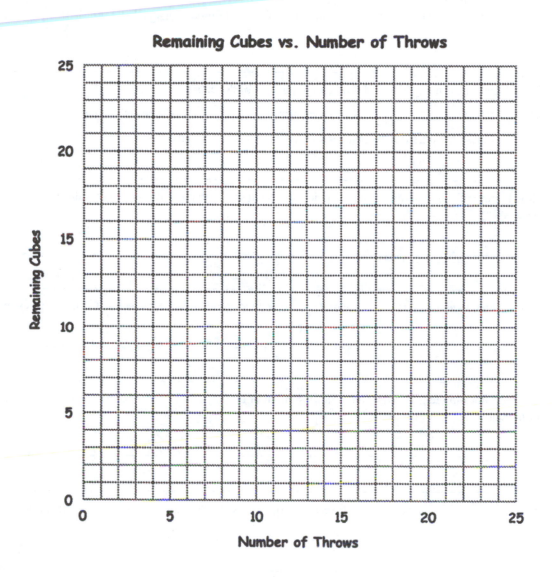

Remaining Cubes vs. Number of Throws

First substance (one side on cube) color or line pattern:

Second substance (two sides on cube) color or line pattern:

Third substance (three sides on cube) color or line pattern:

Get a Half-Life

Summing Up

1. How many rolls did it take for the number of each colored cube to be reduced by half? These are your half-life readings.

   One-side color: _____   Two-side color: _____   Three-side color: _____

2. The half-life of a decaying substance is measured in units of time. What is the unit of half-life used in this simulation?

   _____

3. In each case, how many rolls did it take to remove all of the cubes?

   One-side color: _____   Two-side color: _____   Three-side color: _____

4. Which of these hypothetical substances would be the most radioactive?

   _____

5. How might you simulate the radioactive decay of a substance that decays into a second substance that also decays?

   _____

   _____

6. Is it possible to estimate the half-life of a substance in a single throw? How accurate might this estimate be?

   _____

   _____

7. Are your lines in the graph for Step 8 fairly straight or do they curve? Do these lines correspond to a constant or nonconstant rate of decay?

   _____

   _____

8. a. Substance X has a half-life of 10 years. If you start with 1000 g, how much will be left after:

   i. 10 years? _____

   ii. 20 years? _____

   iii. 50 years? _____

   iv. 100 years? _____

   b. Will this sample of substance X ever totally disappear? If so, estimate how soon. If not, explain.

   _____

   _____

Name _____ Section _____ Date _____

CONCEPTUAL PHYSICS | Activity

Chain Reaction

Purpose
To simulate a simple chain reaction

Apparatus
100 dominoes
large table or floor space
stopwatch

Discussion
Give your cold to two people who in turn give it to two others who in turn do the same on down the line, and before you know it, everyone in class is sneezing. You have set off a chain reaction. This is similar to what happens within a lump of uranium-235 where one neutron triggers the release of two or more neutrons, which trigger the release of even more neutrons. That one neutron has set off a chain reaction. Very quickly, so many neutrons are produced that the lump of uranium explodes as an atomic bomb.

In this activity, you'll explore this idea of the chain reaction.

Procedure
Step 1: Set up a strand of dominoes about half a domino length apart in a straight line. Gently push the first domino over, and measure how long it takes for the entire strand to fall over.

Step 2: Arrange the dominoes again as in the figure, so that when one domino falls, another one or two are toppled over. These topple others in chain-reaction fashion. Set up until you run out of dominoes or table space. When you finish, push the first domino over and measure how long it takes for all the dominoes to fall. Notice the number of the falling dominoes per second at the beginning versus the end.

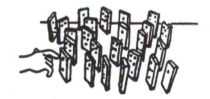

Summing Up
1. Which reaction, wide-spaced or close-spaced dominoes, took a shorter time?

_____

_____

2. How did the number of dominoes being knocked over per second change for each reaction?

_____

_____

3. What caused each reaction to stop?

_____

_____

4. Now imagine that the dominoes are the neutrons released by uranium atoms when they *fission* (split apart). Neutrons from the nucleus of a fissioning uranium atom hit other uranium atoms and cause them to fission. This reaction continues to grow if there are no controls. Such an uncontrolled reaction occurs in a split second and is called a *nuclear explosion.* How is the domino chain reaction similar to the nuclear fission process?

_____

_____

_____

_____

5. How is the domino reaction dissimilar to the nuclear fission process?

_____

_____

_____

_____

Appendix A:
Accuracy, Precision, and Error

These appendices are not definitive, complete references on these subjects. Rather, they are introductions to these topics, intended to serve the needs of this manual. They are simplified and leave much to be discovered in later coursework and practice.

Accuracy and Precision

Suppose the mass of a gold coin were measured using the best scale available. The resulting measurement would be both accurate and precise. Suppose that value were 27.96458 grams. If measured with the best scale available, that measurement would be considered the accepted value.

A second scale, known to be of lesser quality, gives a value of 28 grams. This measurement would best be described as being accurate but not precise. It is close to the accepted value, but has only two significant figures. ***Being close to the accepted value makes it accurate.***

A third scale, again of lesser quality than the original, gives a value of 35.2156 grams. This measurement would best be described as precise but not accurate. It has six significant figures, but isn't very close to the accepted value. ***Having many significant figures makes a value precise.***

A fourth scale, of truly inferior quality, gives a value of 15 grams. This measurement is neither accurate nor precise.

A measurement is considered accurate if it is close to the accepted value. A measurement is considered precise if it includes many significant figures (measures a value across many orders of magnitude).

Random and Systematic Error

One cannot make a measurement without error. There is error in every measurement. Error can be minimized, but not eliminated. In laboratory measurements, there are generally two kinds of error: random and systematic. The meanings of and differences between random and systematic error can be illustrated using an archery analogy.

During target practice, an archer launches 10 arrows at a target. The archer hopes to get all 10 arrows to the center of the target. Consider the results from four archers.

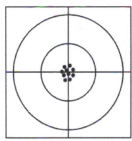 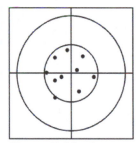 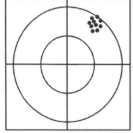 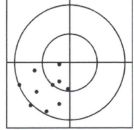

Figure A. Minimal Error   **Figure B.** Mostly Random Error   **Figure C.** Mostly Systematic Error   **Figure D.** Random and Systematic Error

Archer A's shots show very little error. They center on the "bull's eye" ***and*** form a tight cluster.

Archer B's shots suffer from random error. They center on the "bull's eye" but are scattered: some high, some low, some left, and some right. The error seems to be ***different*** with each shot.

Archer C's shots suffer from systematic error. Though the cluster is tight, they do not center on the "bull's eye." Each shot was high and to the right. The error seems to be ***the same*** with each shot.

Archer D's shots suffer from both random and systematic error. They are scattered and do not center on the "bull's eye."

If called upon to describe sources of error in an activity or experiment, be careful about citing *human error.* Students sometimes (understandably) confuse human error with mistakes. While error cannot be avoided, mistakes cannot be accepted. When a mistake is made during a procedure, simply discard the resulting observations and repeat the step in careful accordance with the instructions.

Appendix B:
Uncertainty and Significant Figures

Counting Numbers and Exact Values

Some numbers are exact. The number of eggs in a dozen is 12. There is no uncertainty; the value of 12 is exact. No measurement technology or skill will change the value. Many people have 10 fingers and 10 toes. There are 52 cards in a standard deck. A value need not be a whole number to be exact. The price of a gallon of gasoline is an exact value, even though it is often given with a decimal value, such as 2.659 USD. Exact values are often stated, imposed, counted, or given. They are not measurements.

Uncertainty and Precision in Measurements

All measurements include some degree of uncertainty. This section is devoted to the rules used at the introductory level to respect the uncertainty in measurement.

You might use a meterstick—marked in millimeters—to measure the width of a table and find it to be 76.14 cm. If your meterstick were marked only in centimeters, you might find the width to be 76.1 cm. If you used a better device, you might find the width to be 76.138 235 cm. The 76.1-cm measurement is the least precise and has the greatest uncertainty. The 76.138 235-cm measurement is the most precise and has the least uncertainty.

The precision of a measurement is indicated by the number of significant figures it includes.

76.1 cm	three significant figures
76.14 cm	four significant figures
76.138 235 cm	eight significant figures

Some Rules of Significant Figures

1. **In numbers that do not contain zeros, all the digits are significant.**

   EXAMPLES:

4.132 7	five significant figures
5.14	three significant figures
369	three significant figures

2. **All zeros between significant digits are significant.**

   EXAMPLES:

8.052	four significant figures
7059	four significant figures
306	three significant figures

3. **Zeros to the left of the first nonzero digit serve only to fix the position of the decimal point and are not significant.**

   EXAMPLES:

0.006 8	two significant figures
0.042 7	three significant figures
0.000 350 6	four significant figures

4. **In a number with digits to the right of the decimal point, zeros to the right of the last non-zero digit are significant.**

    EXAMPLES:

53	two significant figures
53.0	three significant figures
53.00	four significant figures
0.002 00	three significant figures
0.700 50	five significant figures

5. **In a number that has no decimal point and that ends in one or more zeros (such as 3600), the zeros that end the number may or may not be significant.**

    The number is ambiguous in terms of significant figures. Before the number of significant figures can be specified, further information is needed about how the number was obtained. If it is a measured number, the zeros are not significant. If the number is a defined or counted number, all the digits are significant.

    Confusion is avoided when numbers are expressed in scientific notation. All digits are taken to be significant when expressed this way.

    EXAMPLES:

4.6×10^{-5}	two significant figures
4.60×10^{-5}	three significant figures
4.600×10^{-5}	four significant figures
2×10^{-5}	one significant figures
3.0×10^{-5}	two significant figures
4.00×10^{-5}	three significant figures

Rounding

Calculators often display eight or more digits. How do you round such a display to, say, three significant figures? Three rules govern the process of deleting unwanted (insignificant) digits from a calculator number.

1. **If the first digit to the right of the last significant figure is less than 5, that digit and all the digits that follow it are simply dropped.**

    EXAMPLE:

    51.234 rounded to three significant figures becomes 51.2.

2. **If the first digit to be dropped is a digit greater than 5, or if it is a 5 followed by a digit other than zero, the excess digits are dropped and the last retained digit is increased in value by one unit.**

    EXAMPLE:

    51.35, 51.359, and 51.359 8 rounded to three significant figures all become 51.4.

3. **If the first digit to be dropped is a 5 not followed by any other digit, or if it is a 5 followed only by zeros, an odd-even rule is applied.**

    That is, if the last retained digit is even, its value is not changed, and the 5 and any zeros that follow are dropped. But if the last digit is odd, its value is increased by one. The intention of this odd-even rule is to average the effects of rounding off.

    EXAMPLES:

    74.250 0 to three significant figures becomes 74.2.

    89.350 0 to three significant figures becomes 89.4.

Appendix C:
Percent Error and Percent Difference

Percent Error

There are times when you want to compare a value found through experiment to an accepted (or standard) value. The absolute difference between the two isn't always useful. For example, an experimental value of gravitational acceleration (9.8 m/s^2) that is off by 3 m/s^2 would be unimpressive. But an experimental value for the speed of light ($299,792,458 \text{ m/s}$) that is off by 3 m/s would be very impressive!

A better measure of error is to compare the error to the accepted value. One way to do this is to calculate percent error. For a given experimental value and accepted value, the percent error is

$$\% \text{Error} = \frac{\text{measured} - \text{accepted}}{\text{accepted}} \times 100.$$

Some prefer to express percent error as a positive value only. If so, simply take the absolute value of the result from the equation above. Ask your instructor for guidance.

Percent Difference

There are occasions when you want to compare two measurements. Since both values are measured, neither can act as the accepted standard. For the percent difference calculation, the difference between the measured values is divided by their **average.**

The percent difference between two values, a and b, can be found by the following equation.

$$\% \text{Difference} = \frac{|a-b|}{(a+b)/2} \times 100. \qquad \text{This simplifies to} \qquad \% \text{Difference} = \left| \frac{a-b}{a+b} \right| \times 200.$$

Percent difference is expressed as a positive value.

Appendix D:
SI Prefixes and Conversion Factors

SI Prefixes

In the metric (SI—International System) system of units, the power of 10 notation can be replaced by prefixes that denote a certain power of 10. For example, kilo denotes 10^3 while micro denotes 10^{-6}. Each prefix also has an abbreviation. The abbreviation for kilo is k and the abbreviation for micro is μ. So 1000 g (one thousand grams) is 10^3 g or 1 kg (one kilogram), while 0.000001 m (one millionth of a meter) is 10^{-6} m or 1 μm (one micrometer). There is an SI prefix for every third power of 10. Here are a few of them.

Prefix	Pronunciation	Abbrev.	Value	Prefix	Pronunciation	Abbrev.	Value
pico	PEE koe	p	10^{-12}	**kilo**	KEE loe	k	10^{3}
nano	NAH noe	n	10^{-9}	**mega**	MEH guh	M	10^{6}
micro	MY kroe	μ	10^{-6}	**giga**	JEE guh*	G	10^{9}
milli	MIH lee	m	10^{-3}	**tera**	TARE uh	T	10^{12}

Some computer users say "GIH guh," but they also tend to spell disc as "disk," McIntosh as "Macintosh," and Googol as "Google." Proficiency with computers and proficiency with language rely on different skill sets.

Some Useful Conversion Factors

This list is by no means comprehensive. It includes several factors that may be useful in one or more of the lab activities.

centimeters and meters	1 cm = 0.01 m	and	1 m = 100 cm
inches and meters	1 in = 0.025 4 m	and	1 m = 39.37 in
feet and meters	1 ft = 0.308 4 m	and	1 m = 3.281 ft
yards and meters	1 yd = 0.914 4 m	and	1 m = 1.094 yd
grams and kilograms	1 g = 0.001 kg	and	1 kg = 1000 g
miles per hour and meters per second	1 mph = 0.447 m/s	and	1 m/s = 2.24 mph
cubic centimeters and cubic meters	$1 \text{ cm}^3 = 0.000\ 001 \text{ m}^3$	and	$1 \text{ m}^3 = 1\ 000\ 000 \text{ cm}^3$

Appendix E:
Graphing

PREPARE THE GRAPH SPACE

1. Use as much of the available graphing space as you can. Leave room for labeling the scale, quantity, and units of each axis.

2. Label the scale of each axis. Label the origin as 0 for both axes. Select a scale so that the range of data fills the graph. Make sure your scales are manageable. Maybe let four squares represent one whole unit of measurement (e.g., 1 meter or 1 volt). But don't make seven squares represent one unit. But again, fill as much of the page as is practical.

3. Label the quantity and units of each axis. Identify the quantity; abbreviate the units parenthetically. "Time (s)," or "x (cm)," for example.

4. Title the graph. You may choose to give the graph a descriptive title, such as "Motion of a Toy Car." But you should always include a title bearing the names of the quantities, such as "Position vs. Time." In such a title, the first quantity listed is the one represented on the vertical axis.

PLOT THE DATA

1. Make a small but visible mark at each data point. If the point (0, 0) is a valid data point, be sure to mark it, too.

2. Resist the urge to connect the dots.

3. When you are done, you will have a so-called "scatter graph."

DRAW A LINE OF BEST FIT FOR A LINEAR PLOT (IF ASKED TO DO SO)

1. A line of best fit is not found by connecting the data point dots. So do **not** connect the dots!

2. If plotting a linear "best-fit" line, make a single straight line that comes as close as possible to all the data points. The line may not pass through any of the actual data points (depending on the amount of scatter in the data). It need not pass through the origin or the final data point either.

3. If multiple plots are to be graphed on a single set of axes, be sure to label each best-fit line to distinguish it from the others.

USE THE LINE OF BEST FIT

1. Once the line of best fit has been drawn, use that line to calculate the slope.

2. Disregard the slope between any two particular data points. The line of best fit is the only line to use for data analysis. It is as if the original data no longer exists, and only the line remains.

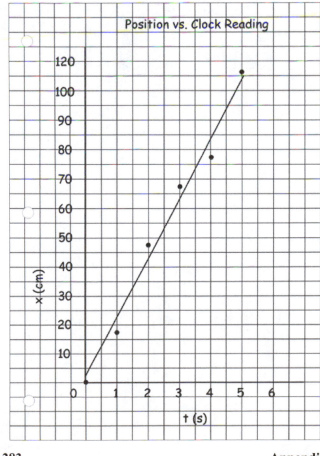